Agricultural Growth for Poverty Alleviation

ABOUT THE BOOK

Agricultural growth has long been recognized as an important instrument for poverty reduction. Yet, measurements of this relationship are still scarce and not always reliable. The authors present additional evidence at both the sectoral and household levels based on recent data. Agriculture Growth Rate in India GDP had been growing earlier but in the last few years it is constantly declining. Still, the Growth Rate of Agriculture in India GDP in the share of the country's GDP remains the biggest economic sector in the country. Agricultural development has three distinct but related dimensions: the physical-technical, the economic-financial, and the institution-human. The physical-technical dimension addresses land utilization, agricultural technologies, research and extension, agricultural inputs, farm-to-market access, productivity and production maximization, and so on. Agricultural development from the economic-financial perspective is concerned with costs, factors of production, terms if trade, pricing policies, subsidies, incentives, credit, return on investment, market mechanisms, and the like. Agriculture dominates non-agricultural activities by its potential to reduce poverty, whether the comparison is within or between countries. More than half of the reduction in poverty achieved in the selected countries during the years under study can be attributed to growth in agricultural incomes. The book basically contains basic concepts and ideas on various courses purely meant for teaching undergraduates and some practical examples are also included in the book.

ABOUT THE AUTHOR

Rita Mishra, M.Sc. Agriculture and pursuing her Ph.D. She is presently working in Nalanda College, Bihar as Professor, she is an erudite scholar and seasoned teacher to the core is highly qualified and teaching the subject at a reputed civil service Academy based in Delhi for last seven years .She has attended a number of seminars and conferences and also organised several workshops for civil service students and published several articles in various reputed journals pertaining to problems of farmers, future of agriculture etc. She published more than 13 books on Agriculture fields.

Agricultural Growth for Poverty Alleviation

Rita Mishra

AMIGA PRESS. INC
Delhi -110051

AMIGA PRESS INC.

H.O :- 28-A, Indra Park,

Chander Nagar, Delhi :- 110051 (India)

Tel :- 011-22504028, +91-98736-57000

Email :- amigapress@gmail.com
amigaeditorial@gmail.com
amigapub@gmail.com

Website :- www.amigapress.co.in

This edition is published by Amiga Press Inc.

ISBN : 978-93-87295-38-4

First Published 2019

Printed and bound by Replika Press Pvt. Ltd.

Preface

Agriculture Growth Rate in India GDP had been growing earlier but in the last few years it is constantly declining. Still, the Growth Rate of Agriculture in India GDP in the share of the country's GDP remains the biggest economic sector in the country.

Agricultural diversification is an important mechanism for economic growth. It depends, however, on there being opportunities for diversification and on farmers' responsiveness to those opportunities. Agricultural diversification can be facilitated by technological breaks-through, by changes in consumer demand or in government policy or in trade arrangements, and by development of irrigation, roads, and other infrastructures. Conversely, it can be impeded by risks in markets and prices and in crop-management practices, by degradation of natural resources, and by conflicting socio-economic requirements-perhaps for employment generation, or for self-sufficiency or foreign-exchange-earning capacity in particular crops or livestock or fishery or forest products.

Agricultural productivity can therefore be seen as a first step or engine of growth leading to greater income for a country. It is interesting to note that historically no poor countries have reduced poverty only through agriculture, but almost none have achieved it without increasing agricultural productivity in the first instance. Agricultural growth is an essential complement to growth in other sectors.

Urban farming is generally practiced for income-earning or food-producing activities though in some communities the main impetus is recreation and relaxation. Urban agriculture contributes to food security and food safety in two ways: first, it increases the amount of food available to people living in cities, and, second, it allows fresh vegetables and fruits and meat products to be made available to urban consumers. A common and efficient form of urban agriculture is the biointensive method. Because urban agriculture promotes energy-saving local food production, urban and peri-urban agriculture are generally seen as sustainable practices.

The city seeks green spaces emanating vitality and beauty for enriching existing spaces and responding to the citizens' needs for a new distribution of urban spaces and better quality of life. That is why it is important to organize the different spaces by ensuring their individuality and their unique formal, dimensional and qualitative differences. The Green System is based on the principle of linkages and continuity of the green spaces to ensure uses and functions according to their locations. Green spaces link urban networks and have a very important meaning for the city's image and operations. In order to address urban agriculture as part of the Green System, it is necessary to define its roles within this system in a flexible, easily implemented work plan. As part of the Green System, urban agriculture will be able to have an integrated set of activities and reach its full potential, with the flexibility to include new elements in the system.

The yields for various crops in these states differ greatly. While Tamil Nadu had the highest yield in rice, oil seeds and sugarcane, Punjab enjoyed the highest yields in wheat, coarse cereals, pulses and food grains. Karnataka on the other hand is seen to do well in cotton and Bihar performed quite well in pulses and coarse cereals. Further analysis and findings by Kalirajan and others (2001) show that Punjab had made remarkable achievements on the agricultural front while Bihar had remained stagnant in the last two decades, with Karnataka and Tamil Nadu showing moderate achievement. Clearly, differences in physical endowments, climatic conditions and institutional characteristics are some of the reasons for the varying productivity performance. Thus, having across the board economic reforms is likely to work less effectively than state-specific policy measures that enable each state's agricultural yields to reach their full potential. The comparative advantage of each state's agricultural production should be determined and with inter-state restrictions removed, total agricultural output would see a very significant increase.

This is a reference book. All the matter is just compiled and edited in nature, taken from the various sources which are in public domain.

The book basically contains basic concepts and ideas on various courses purely meant for teaching undergraduates and some practical examples are also included in the book.

—Editor

Contents

1

Agriculture Diversification and Poverty Alleviation

Agriculture Growth Rate in India GDP had been growing earlier but in the last few years it is constantly declining. Still, the Growth Rate of Agriculture in India GDP in the share of the country's GDP remains the biggest economic sector in the country.

In the 1980s, Indian policymakers shifted their focus from food self sufficiency to generating additional income in rural areas as a means of tackling the problem of poverty, which was concentrated in rural areas. Acceleration of agricultural growth, with a special focus on improving the position of small farmers and extending the productivity revolution to non-irrigated areas was seen as a critical part of the strategy for poverty alleviation. This effort was supplemented targeted anti poverty programs to address the needs of vulnerable groups who may not benefit sufficiently from general agricultural growth. India achieved considerable success with this approach in the 1980s. Growth of agricultural gross domestic product (GDP) accelerated to about 4.7 percent in the 1980s, compared with only 1.4 percent in the 1970s. This, agricultural growth, together with the beginning of economic reforms in the nonagricultural sector, pushed up the growth rate of overall GDP to around 5.8 percent in the period 1980-81 to 1989-90 compared with about 3 percent in the 1970s.

India's growth was disrupted at the start of the 1990s by a major balance of payments crisis which led to the adoption of an extensive process of structural reforms. It took time to regain momentum and it was only in 1993-94 that the economy got back on track, clocking an average growth rate of 6.8 percent in the three years 1993-94 to 1995-96. This acceleration in growth in the post reform period led policymakers to set a more ambitious GDP growth target of 8 percent a year for the Ninth Plan period (1997-98 to 2001-

2002), to be supported by a growth rate of 4 percent a year in agriculture. The projected growth of 4 percent per year in agriculture was clearly inline with the average growth of 3.8 percent achieved in the period 1990-91 to 1996-97.

However, actual performance since the mid 1990s has been disappointing. Agricultural growth slowed to 2 percent a year in the Ninth Plan period, and overall economic growth was only 5.5 percent, well below the 8 percent target. Since agriculture accounted for about 25 percent of GDP, the shortfall of more than 2 percentage points in agricultural GDP growth compared with the target accounts directly for a shortfall of about half a percentage point in GDP growth. If the indirect effects of more rapid agricultural growth on other sectors are taken into account, the total impact on GDP growth may have been as much as one percentage point.

These shortfalls were known, when the Tenth Plan (covering the period 2002-03 to 2006-07) was formulated, but it was assumed that the poor performance of agriculture was due to temporary factors such as poor monsoons and depressed agricultural commodity prices in world markets following the East Asian meltdown. The Tenth Plan therefore adopted the same targets of 8 percent growth in GDP and 4 percent growth in agriculture. Experience in the first three years of the Tenth Plan period has sounded some alarm bells. GDP growth has averaged about 6.5 percent, but agricultural GDP in these years (2002-03 to 2004-05) has grown by only 1.1 percent per year. The loss of dynamism in agriculture explains most of the shortfall in aggregate GDP growth.

Slower growth in agriculture also has direct implications for poverty reduction in rural areas. Official figures suggest that the incidence of poverty fell from 36 percent in 1993-94 to 26 percent in 1999-2000. The comparability of these numbers has been questioned because of recent changes (ostensibly improvements) in the methods for measuring consumption in household surveys, but there is a broad consensus that if corrections are made to ensure comparability, the percentage of the population in poverty has declined significantly, though less than in the official figures. However, even the official figures show less decline than what had been targeted, and this is undoubtedly a reflection of the slowdown in agricultural growth. Slow growth in agriculture is also at the root of growing evidence of distress in the farming community. Surveys show that a large percentage of farmers want to leave farming because they find it is no longer sufficiently profitable. The uncertainty associated with farming has also increased and farmers lack effective means of insuring against such risks. There are larger market uncertainties associated with new crops and poultry because of greater vulnerability because of falling ground water levels. There is evidence of increased indebtedness arising from the inability to cope with risks.

Recognizing these problems, the Government has undertaken a comprehensive review of the strategy for agriculture in order to come up with a new deal for agriculture and the rural economy in general. Remedial action will be needed on several fronts including increased public investment in irrigation and rural roads, better management of existing irrigation systems and of water resources in dry land areas, a strengthened agricultural research system and more effective extension, improvements in the production and distribution of certified seeds, improvements in the credit delivery system, and innovative steps in marketing and contract farming to support the diversification of Indian agriculture.

Irrigation

Water is a critical constraint on raising agricultural productivity and much of the success of the Green Revolution came from improved productivity in areas of assured irrigation provided through canals or (much more significant) through ground water utilization. The scope for expanding irrigation through large and medium scale projects has yet to be fully exploited. Out of the total of 59 million hectares that could be irrigated through such projects only 40 million hectares have been irrigated. The slow pace of exploitation of irrigation potential is due to lack of resources in state governments and the tendency to spread available resources thinly over too many projects. Additional public investments in this area are therefore essential for early utilization of the potential.

Effective maintenance of the existing system of canal irrigation also suffers because the irrigation departments of the states lack resources. This in turn is because water charges are kept too low, covering only 20-25% of the operations and maintenance cost of the system in most states. Poor maintenance leads to loss of water through seepage, with the result that water use efficiency is very low – around 25 to 40 percent instead of 65 percent that should be attainable. Low water charges also encourage highly water intensive crops at the upper end of the canal network, leaving tail-end portions starved of water.

The solution lies in rationalization of water rates to ensure adequate financial resources to cover maintenance and resort to participatory irrigation management to give farmers a stake in the operation and maintenance of the system. Some interesting experiments in these have promise. Maharasthra has recently established a Water Regulatory Authority to set water charges in a non-political manner. Several states are also experimenting with involving water user associations (WUAs) in the operation of the canal systems. Ideally the WUAs should be empowered to collect water charges and to retain a portion of the collection to maintain the portion of the distribution network operating in their area.

Ground water utilization played a major role in expanding irrigation in the 1980s but uncontrolled exploitation of groundwater has led to serious depletion of the water table in many parts of the country. Overexploitation is encouraged by the policy of massive under pricing of electricity for agricultural use, with a few states having made electricity for farmers completely free. Even where it is not free, the charge for electricity is a fraction of the average cost, and is not based on metered use. Instead there is a fixed charge for presumed usage, based on the capacity of the pump, an arrangement which implies that the marginal cost of electricity for pumping ground water is zero. Under pricing canal water and electricity are clearly highly distortionary, given the need to conserve water use. They are also distributionally unfair because the benefits of under priced water accrue disproportionately to upper end farmers whereas under priced power enables those able to afford larger pumps to lower the water table denying water to farmers who can only afford shallow wells.

The investment requirements of irrigation are massive. Completion of all unfinished projects alone is estimated to cost approximately US$ 20 billion. In addition, provision must be made for new irrigation projects (large, medium and small), which together will require about US$ 45 billion. The total requirement is therefore about US$ 65 billion.

POVERTY TARGETING PROGRAMS IN INDIA: AN OVERVIEW

Despite severe fiscal imbalances in the country, manifested in continued high fiscal deficits through most of the 1990s, CSS have proliferated with abandon during the decade. In the terminal year of the Ninth 5-year Plan (2001), there were 360 schemes in operation as CSS. The latest 5-year Plan in the country has called for a convergence of similar schemes and the elimination of schemes that have outlived their utility, viewing the "mushrooming growth" of CSS as a "case of the state overreaching itself". The Planning Commission recommended eliminating 48 schemes, merging 161 schemes into 53, and retaining the remaining 135 schemes, implying a continuation forward of a total of 135 schemes.

The large number of schemes under CSS are a major source of ambiguity in assessing total government expenditures on PTPs, since some of the schemes are directly targeted at poverty alleviation while others have less direct yet substantial benefits for the poor in the medium and long term. The selection of specific schemes as poverty targeted will necessarily be qualitative, and vary according to sources. As can be seen, expenditures have increased substantially in nominal terms, by almost a factor of 500 percent. However, due to relatively high inflation rates in the first half of 1990s, the increase in real terms – in 1993-94 prices – has been relatively more modest. In particular, expenditures in real terms remained relatively static during the 1990s following an increase in 1993-94, and have only increased more recently in 2000.

Table: Poverty programs in India, 1999-2000

Name of the Program/Ministry	Budget allocation in 1999-00 (Rs billion).
Rural Development Schemes	94.3
Food Subsidy	92.0
Subsidy on kerosene	80.4
Health & Family Welfare(only 70% of the outlay)	28.4
Social Justice & Empowerment Sector	12.1
Integrated Child Development Services	11.5
Mid day meal	10.3
DPEP	7.6
Watershed development through agriculture	2.3
Tribal Development	1.9
Swarnajayanti Shahari Rozgar Yojana (Urban Poverty)	1.8
Total	342.6

Source: GoI (2000), ch. 31

For comparison, table provides estimates by the Planning Commission on poverty related schemes in 1999-2000. According to these estimates, total expenditure on poverty programs was Rs. 342.6 billion, but if we exclude the subsidies on food and kerosene oil, the total is only Rs. 170.2 billion. However, these data do not include transfers directly to the district governments by the centre, that as already noted can be substantial (Rs. 150 billion in 2002).

Although the estimates vary, they are quantitatively in the same order of magnitude. Nayak et.al. (2003) estimated total expenditures on schemes under CSS at approximately Rs. 250 billion in 2000, including direct transfers from centre to the districts. This amount was almost 3-4 times higher than all Official Development Assistance to India in 2000, which was USD 1.49

billion or almost Rs. 70 billion (at an exchange rate of Rs. 47/USD 1). Thus, notwithstanding a relatively static trend in real terms through much of the 1990s, expenditures by Government of India on PTPs are higher by significant orders of magnitude compared to all ODA coming into the country. Not all CSS are narrowly defined poverty-targeted programs, since some of them may be more broadly targeted, focusing on irrigation or road development for example. At the same time, these amounts are also supplemented by expenditures made by state governments to share in costs of the schemes under CSS.

The relative composition of schemes under CSS over the last two decades in terms of broad heads of development. Evidently, the share of schemes under agriculture and rural development and social sectors has been rising consistently, exceeding 60 percent in the previous decade, at the expense of schemes targeted at industry and minerals, energy and communications sectors. There was a marked increase in 2002-03, the first year of the Tenth 5 year plan, in the share of schemes directed at transport and in the share of "others", which is due to several new schemes announced for impoverished North-Eastern states of the country. The increased share of transport reflects a major expansion of road construction in India funded by the centre but implemented by states.

The large expenditures on poverty reduction through the CSS are difficult to track for two reasons. First, they are routed through different ministries of the Government of India with little centralization of the relevant information. For example, although the aggregate budget of theGovernment of India provides budgetary allocations on different schemes, the information is scattered across accounts of different ministries implementing the schemes. In addition, even within the relevant ministries, the funds are allocated across numerous schemes, some large and some quite miniscule. As noted already, while the expenditures on the CSS in real terms have not risen sharply, the schemes themselves have proliferated resulting in numerous instances of renaming schemes accompanied by merging and restructuring of schemes that allocate specific components into other newly created/renamed schemes. The result is erosion of transparency.

An overview of direct poverty targeted programs in India, identifying major schemes under the CSS and the ministries implementing the schemes. Only schemes with central funding exceeding Rs 1 billion in 2001-02. Clearly, several ministries of the central government are involved in implementing PTPs, but the major entity involved is the Ministry of Rural Development, in terms of number of major schemes. This is natural given the vast majority of poor in India live in rural areas. Allocations are much higher for schemes implemented by the Department of Public Distribution under the Ministry of Consumer Affairs, that provides subsidized food under the PDS targeted to those below the poverty line, and the Ministry of Fertilizers. However, fertilizer subsidies are distributed to the producers rather than directed to poor farmers.

AGRICULTURAL BIOTECHNOLOGY IN POVERTY ELEVATION

At the dawn of the twenty-first century, about 900 million or 68 percent of the world's poor people live in Asia; about 500 million in South Asia, 300 million in East Asia, and 100 million in Southeast Asia and the Pacific (World Bank 1999). In addition, about 526 million people, including 160 million children, are undernourished2 (FAO 1999c). Not only do they lack access to sufficient money to buy food and other essentials, but neither do they have access to sufficient schooling, adequate housing, nor medical care. Those in rural areas are often short of water and fuel. Fertile land and water for farming are increasingly scarce. Urban poor lack money to buy enough food. That which they can afford may be deficient in protein and essential vitamins and minerals.

Although the absolute numbers of people living in poverty in Asia today are unacceptable, the situation could be much worse. In 1970, 60 percent of all Asians lived in poverty. That figure has been cut by almost half, with about one third of all Asians living in poverty in 2000. Also, countries such as Bangladesh, the People's Republic of China (PRC), and India have moved from periodic famines to almost self-sufficiency in food production. However, further efforts are needed to reduce poverty by another 50 percent by 2015, as targeted by world leaders during the World Food Summit in 1997. The latter half of the twentieth century saw impressive advances in science and technology. We now have the capacity to apply this knowledge to reduce poverty and improve food security.

TARGETED ANTI-POVERTY PROGRAMS

While efforts to increase agricultural productivity and thereby increase farm incomes and employment are a major instrument for poverty alleviation, they will need to be supplemented by special targeted program aimed at improving the welfare of vulnerable groups in rural areas.

Employment programs in rural areas have been the most important of these anti poverty programs and India has a long history of such programs.

Building on this tradition, a Rural Employment Guarantee Act has been enacted which provides assurance of up to 100 days of employment at the minimum wage to each household in rural areas wishing to make use of it.

The employment would be provided on projects chosen by the elected village councils and the guidelines specify that top priority should be given to irrigation and water management schemes. Unlike earlier employment programs, this scheme includes a guarantee in the sense that if employment cannot be provided, unemployment compensation will be provided at least 25 percent of the wage. Although the program open to each household, actual demand for employment is expected to be limited to households below the poverty line. The coverage of the Act will initially be implemented to 200 of the most backward districts (about one-third of the total districts in the country). Together with other special programs relating to provision of housing for the poor, old age insurance, and schemes for supporting self employment, this program will provide an element of social security that should help to reduce poverty.

REDUCING POVERTY THROUGH AGRICULTURAL SCIENCE

Science and technology underpinned the economic and social gains in Asia over the past 30 years. In agriculture, these gains came to be known as the Green Revolution. Between 1970 and 1995, cereal produc-tion in Asia doubled, calorie availability per person increased by 24 percent, and real food prices halved (IFPRI 1997). Although the region's population grew by 1 billion people, overall food production more than kept pace with population growth (McCalla 1998). These food production increases were achieved largely by the cultivation of high-yielding varieties (HYVs) of rice and wheat, accompanied by expansion of irrigated areas, increases in fertilizer and pesticide use, and greater availability of credit.

The scientific basis for the Green Revolution stemmed from national and international research programs that led to the development and distribution of new HYVs, particularly of rice and wheat. The first generation of these new varieties were based on the introduction of new genes for dwarfing that made the HYVs shorter, more responsive to fertilizers, and less prone to falling over or lodging when fertilized and irrigated. Subsequent varieties also carried genes that gave increased pest and disease resistance and improved taste and grain quality.

The key elements in improving food security in Asia from 1970-95 were government policies reflecting a belief that investments in increasing agricultural productivity were a prerequisite to economic development. These national policies were supported by political leaders in Asia and by both the public and private sectors of the international community. This mix of supportive public policies, scientific discoveries, and public and private investments in rural Asia, particularly in irrigation, credit, and inputs, led to substantial reductions in poverty and improved food security throughout Asia over the past 30 years. Increased agricultural productivity, rapid industrial growth, and expansion of the nonfarm rural economy have all contributed to almost a tripling of per capita gross domestic product across Asia since 1970 (ADB 2000b, Pinstrup-Andersen and Cohen 2000).

PRESENT PROBLEMS

Despite these successes, problems remain. The intensification of agriculture and the reliance on irrigation and chemical inputs has led to environmental degradation, increased salinity, and pesticide misuse. Deforestation, overgrazing, and overfishing also threaten the sustainable use of natural resources.

Green Revolution technologies had little impact on the millions of smallholders living in rainfed and marginal areas, where poverty is concentrated. Furthermore, the Green Revolution has already run its course in much of Asia. Wheat and rice yields in the major growing areas of Asia have been stagnant or declining for the past decade, while population continues to increase (Pingali et al. 1997). The key lessons learned from the Green Revolution are: (i) it has benefited farmers in irrigated areas much more than farmers in rainfed areas thus worsening the income disparity between the two groups, (ii) it overlooked the rights of women to also benefit from the technological advances, and (iii) it promoted an excessive of use of pesticides that are harmful to the environment. As countries became self sufficient in food, government investments declined in the agricultural sector and in science and technology across the region. This reflects a worldwide trend toward declining public investments in the rural sector and in agricultural research and development (R&D), nationally and internationally.

In Asia, private sector investments in the rural sector and related R&D have concentrated on export commodities. The downward trends in public investments by governments and development agencies in smallholder agriculture over the past decade have not been matched by a concomitant rise in private investments. Similarly, there is little (and few incentives for) private R&D on the food crops, livestock, fisheries, and aquaculture systems important for food security and poverty reduction in rural Asia.

RATIONALE FOR ADB'S INVOLVEMENT IN BIOTECHNOLOGY

Biotechnology has the potential to (i) increase crop and animal productivity; (ii) improve nutritional quality; (iii) broaden tolerance of crops for drought, salinity, and other abiotic stresses; and (iv) increase resistance of crops to pests and diseases. These potential benefits will have significant impact in increasing food production and reducing poverty in DMCs if they can be applied to problems of the poor farmers in the tropics. As with any technology, biotechnology brings with it potential risks. To maximize the benefits and minimize the risks, the introduction and use of biotechnology in DMCs or elsewhere must be thoughtfully managed by the public and private sectors. The key risks that relate to the application of new developments in biotechnology for the public good are food and environmental safety, economic concentration, and intellectual property (IP) management.

To explore the opportunities and risks of biotechnology, ADB has decided to undertake a comprehensive study on the use of biotechnology to reduce poverty and achieve food security in cooperation with the Australian Centre for International Agricultural Research (ACIAR)/ Australian Agency for Agricultural Development (AusAID). There are many reasons to support this initiative:

(i) As a premiere development institution, ADB should be well informed on the development of biotechnology and the issues surrounding modern biotechnology.

(ii) Biotechnology has been identified by the Rural Asia Study as one of the emerging challenges in Asia (ADB 2000b).

(iii) Biotechnology has been identified by the Consultative Group on International Agricultural Research (CGIAR) as a powerful tool to fight poverty when used in conjunction with other agricultural research (Persley and Lantin 2000).

(iv) ADB is responsible for assisting the DMCs in dealing with potential risks of biotechnology and providing information on various biotechnology issues.

(v) ADB's Policy on Agriculture and Natural Resources Research recommends that ADB support research to increase farm productivity through modern biotechnology approaches, while giving appropriate consideration to indigenous and traditional knowledge (ADB 1995).

ADB's Poverty Reduction Strategy

While the need for further intensification of agricultural production in Asia is clear, intensification strategies must change to avoid adverse environmental impact and to reverse the effects of earlier practices. Enhanced, but inefficient use of irrigation and mineral fertilizers over the past three decades has had negative side effects such as soil salinity and nutrient leaching. With crop intensification, incidences of pests and diseases also increased (Pinstrup-Andersen and Cohen 2000). The following strategies are needed to meet the food demand in Asia over the next 25 years.

(i) Sustainable productivity increases in food, feed, and fiber crops.

(ii) Reducing chemical inputs of fertilizers and pesticides and replacing these with biologically based products.

(iii) Integrating soil, water, and nutrient management.

(iv) Improving the nutrition and productivity of livestock and controlling livestock diseases.

(v) Sustainable increases in livestock, fisheries, and aquaculture production.

(vi) Increasing trade and competitiveness in global markets.

As rural poverty persists in Asia, agriculture will play a prominent role in achieving equitable and sustainable rural growth in the twenty-first century. Even when rural people do not work directly in agriculture, they rely on nonfarm employment and income closely related to agriculture. Where there are large numbers of rural poor, agricultural growth is a catalyst for broad-based economic growth and development. Agriculture's linkages to the nonfarm economy generate employment, income, and growth in the rest of the economy. A healthy agricultural economy also offers incentives for natural resource conservation (Pinstrup-Andersen and Cohen 2000).

The ADB Poverty Reduction Strategy (ADB 1999) sees three factors, pro-poor, sustainable economic growth; good governance; and social development as key elements for reducing poverty. Biotechnology may contribute toward achieving the poverty reduction goals in several components of the strategy:

(i) Increasing priority for agriculture and rural development.

(ii) Addressing environmental considerations.

(iii) Increasing the public and private sector roles in poverty reduction.

(iv) Encouraging regional and subregional cooperation.

(v) Coordinating efforts of funding agencies.

The challenge is how to use new scientific developments such as biotechnology, together with information and communications technology, to make the complex agricultural systems of Asia more productive and sustainable. Good governance is also crucial to ensure that new agricultural biotechnology reaches the poor.

SCIENTIFIC CONTEXT

Promethean Science

The pace of change in modern science has led some to term it Promethean science, acknowledging both its risks and benefits (Serageldin and Persley 2000). Modern science encompasses new developments in the biological, physical, and social sciences. In biology, discoveries over the past 20 years allow the better understanding of the structure and function of human, animal, and plant genes. At the same time, new discoveries in the physical sciences underpin the revolution in information and communications technologies. Geographic information systems enable characterization of agro-ecosystems and offer means by which new technologies can be customized to the needs of particular agro-ecosystems. The biological and physical sciences also interact in new ways. For example, the ability to analyze large volumes of data is a critical component of various genome projects that are mapping all the genes in an organism, as in the Human Genome Project.

New developments in the social sciences underpin community participation in technology development and evaluation (sometimes termed agro-ecological methods). Participatory methods developed in the social sciences can help in understanding problems and the researchable issues, particularly those of small farmers operating in marginal environments. They may also be used to clarify the concerns of rural and urban dwellers in regard to the deployment of new technologies, including the products of biotechnology. Integration of all branches of modern science and traditional knowledge is required to develop knowledge-intensive solutions to the problems of rural Asia. These solutions need not be only technically feasible but also socially acceptable. Indeed, the potential value of modern science to agriculture and the environment in Asia will require the efforts of all stakeholders, including civil society, farmer cooperatives, producers, consumers, governments, and development agencies.

Definition, History, and Scope of Biotechnology

Biotechnology, broadly defined, includes any technique that uses living organisms, or parts of such organisms, to make or modify products, to improve plants or animals, or to develop microorganisms for specific use. It ranges from traditional biotechnology to the most advanced modern biotechnology. Biotechnology is not a separate science but rather a mix of disciplines (genetics, molecular biology, biochemistry, embryology, and cell biology) converted into productive processes by linking them with such practical disciplines as chemical engineering, information technology, and robotics. Modern biotechnology should be seen as an integration of new techniques with the well-established approaches of traditional biotechnology such as plant and animal breeding, food production, fermentation products and processes, and production of pharmaceuticals and fertilizers (Doyle and Persley 1996).

The key components of modern biotechnology are listed below.

(i) Genomics: The molecular characterization of all genes in a species.

(ii) Bioinformatics: The assembly of data from genomic analysis into accessible forms, involving the application of information technology to analyze and manage large data sets resulting from gene sequencing or related techniques.

(iii) Transformation: The introduction of one or more genes conferring potentially useful traits into plants, livestock, fish and tree species.

(iv) Genetically improved organism.

(v) Genetically modified organism (GMO).

(vi) Living modified organism (LMO).

(vii) Molecular breeding: Identification and evaluation of useful traits in breeding programs by the use of marker-assisted selection (MAS);

(viii) Diagnostics: The use of molecular characterization to provide more accurate and quicker identification of pathogens; and

(ix) Vaccine technology: The use of modern immunology to develop recombinant deoxyribonucleic acid (rDNA) vaccines for improved control of livestock and fish diseases (Doyle and Persley 1999).

Biotechnology consists of a gradient of technologies, ranging from the long-established and widely used techniques of traditional biotechnology to novel and continuously evolving modem biotechnology techniques. During the 1970s scientists developed new methods for precise recombination of portions of deoxyribonucleic acid (DNA), the biochemical material in all living cells that governs inherited characteristics, and for transferring portions of DNA from one organism to another. This set of enabling techniques is referred to as rDNA technology or genetic engineering.

Modern biotechnology presently includes the various uses of new techniques of rDNA technology, monoclonal and polyclonal antibodies, and new cell and tissue culture methods. Over the past two decades the number of significant advances in modern biotechnology for understanding and modifying the genetics of living organisms has increased dramatically. That has led to greatly increased interest and investment in biotechnology, and increasing concerns as to the power of the new technologies and their safety.

Economic Concentration in Agricultural Biotechnology

Modern biotechnology R&D has been conducted in an institutional and economic environment that differs significantly from the development of the earlier Green Revolution technologies. While the latter were essentially the prerogative of public research institutions and philanthropic foundations, the application of modern biotechnology to agriculture is essentially a competitive, commercial endeavor in which powerful private sector interests compete. Similarly, while the Green Revolution technologies were essentially dedicated to the public, the strengthening and extension of IP protection, particularly since the conclusion of the Uruguay Round of trade negotiations has increased the private character of biotechnologies. Multinational companies in the seed, agricultural chemical, pharmaceutical, and food-processing industries play a major role in biotechnology research. They have invested heavily in in-house research facilities, commissioned research, taken equity positions in new biotechnology firms, and entered into contractual arrangements with public research institutions or universities. As a result of mergers and acquisitions in the past few years, the development of new biotechnology applications in agriculture has become increasingly concentrated in the hands of a decreasing number of companies. The dominant companies that operate within global markets are Aventis, AgrEvo, Dow, DuPont, Monsanto, and Syngenta.

Table: The Evolution of the Science of Genetics, Leading to Modern Biotechnology

1866 Mendel postulates a set of rules to explain the inheritance of biological characteristics in living organisms.

1900 Mendelian law rediscovered after independent experimental evidence confirms Mendel's basic principles.

1903 Sutton postulates that genes are located on chromosomes.

1910 Morgan's experiments prove genes are located on chromosomes.

1911 Johannsen devises the term "gene", and distinguishes genotypes (determined by genetic composition) and phenotypes (influenced by environment).

1922 Morgan and colleagues develop gene mapping techniques and prepare gene map of fruit fly chromosomes, ultimately containing over 2000 genes.

1944 Avery, MacLeod and McCarty demonstrated that genes are composed of DNA rather than protein.

1952 Hershey and Chase confirm role of DNA as the basic genetic material.

1953 Watson and Crick discover the double-helix structure of DNA.

1960 Genetic code deciphered.

1971 Cohen and Boyer develop initial techniques for rDNA technology, to allow transfer of genetic material from one organism to another.

1973 First gene (for insulin production) cloned, using rDNA technology.

1974 First expression in bacteria of a gene cloned from a different species.

1976 First new biotechnology firm established to exploit rDNA technology (Genentech in USA).

1980 USA Supreme Court rules that microorganisms can be patented under existing law (Diamond v. Chakrabarty).

1982 First rDNA animal vaccine approved for sale in Europe (colibacillosis).

First rDNA pharmaceutical (insulin) approved for sale in USA and UK.

First successful transfer of a gene from one animal species to another (a transgenic mouse carrying the gene for rat growth hormone).

First transgenic plant produced, using an agrobacterium transformation system.

1983 First successful transfer of a plant gene from one species to another.

1985 US Patent Office extends patent protection to genetically engineered plants.

1986 Transgenic pigs produced carrying the gene for human growth hormone.

1987 First field trials in USA of transgenic plants (tomatoes with a gene for insect resistance).

First field trials in USA of genetically engineered microorganism.

1988 US Patent Office extends patent protection to genetically engineered animals.

First GMO approved. Human genome mapping project initiated.

1989 Plant genome mapping projects (for cereals and *Arabidopsis*) initiated.

2000 Plant genome mapping projects for rice and Arabidopsis completed, and about 44 million hectares of land planted to GMO crops.

DNA = deoxyribonucleic acid, GMO = genetically modified organism, rDNA = recombinant

DNA, UK = United Kingdom, USA = United States of America.

Source: Adapted from Persley (1990).

Biotechnology R&D has been concentrated in a limited number of industrialized countries, with the United States (US) in the lead in financial and human resources. A growing number of developing countries have invested in biotechnology R&D, but the amounts are small compared to the sums invested by private companies in the industrial world. While private sector investment in agricultural research in general is increasing in developed countries, there is still little private sector biotechnology research effort in developing countries, particularly in Asia (Pinstrup-Andersen and Cohen 2000).

The commercialization and distribution of new agricultural biotechnology products, particularly transgenic crops, is also concentrated in Organisation for Economic Co-operation and Development (OECD) member countries, with a few exceptions (James 2000). These products are for the most part crops of economic importance in industrial country agricultural production and in world trade in agricultural commodities, mainly soybean, maize, cotton, and canola. During the past five years, the area under GMOs3 has increased rapidly from 1.7 million ha to 44.4 million ha, 75 percent of which are in the US. The remaining 25 percent are distributed in both developed and developing countries, including Argentina, Australia, Bulgaria, Canada, PRC, France, Germany, Mexico, Portugal, Romania, South Africa, Spain, Ukraine, and Uruguay.

Relatively little biotechnology research is being undertaken on the problems of small farmers in rainfed and marginal lands. Neither is there much interest in Asia's basic food crops: rice, tropical maize, wheat, sorghum, millet, banana, cassava, groundnut, oilseed, potato, sweetpotato, and soybean. These are considered *orphan crops* because of the private sector's reluctance to work on them. That focus is unlikely to change because of the perception that investments in such orphan crops and from working on problems of small farmers yield limited returns. To participate more fully in the biotechnology revolution, Asian governments will need to expand their capacities to undertake biotechnology research linked to the problems of small farmers and orphan crops. In certain situations, however, there may be opportunities to purchase, license, or import technology applicable in Asia.

Applications of Biotechnology

The applications developed from the new methods in biotechnology place them within the continuum of techniques used throughout human history in industry, agriculture, and food processing. Thus, while modern biotechnology provides powerful new tools, they are used to generate products that fill similar roles to those produced with more traditional methods.

There is now increasing use of modern molecular genetics for genetic mapping and MAS as aids to more precise and rapid development of new strains of improved crops, livestock, fish, and trees. Other biotechnology applications such as tissue culture and micropropagation are being used for the rapid multiplication of horticultural crops and trees. New diagnostics and vaccines are being widely adopted for the diagnosis, prevention, and control of fish and livestock diseases.

The science of genomics (the molecular characterization of all the genes in a species) has dramatically increased knowledge of plant genes and their functions. The new technologies enable greatly increased efficiency of selection for useful genes, based on knowledge of the biology of the organism and the role of specific genes in regulating particular traits. This will enable more precise selection of improved strains. These techniques may be used for more efficient selection in conventional breeding programs. They may also be used for the identification of genes suitable for use in the development of transgenic crops. Thus far, scientists have completed genomic study on rice through the cooperative efforts of several international and private sector institutions led by Japan.

Modem biotechnology permits increased precision in the use of new techniques and a shorter time to produce results. For example, plant breeders and molecular biologists can collaborate to transfer to a highly developed crop variety one or two specific genes to impart a new character such as a specific kind of pest resistance.

New techniques of modern biotechnology accelerate plant and animal breeding. They offer possible solutions to previously intractable problems and difficult targets such as drought tolerance, and enable the development of new products. These products may include more nutritious food, crop varieties with improved tolerance for pests and diseases, and animal vaccines.

It is important to provide appropriate regulatory mechanisms to ensure that products produced by modern biotechnology are as safe as the products of traditional biotechnology. That is especially so when the products are GMOs that might interact with the environment. Our knowledge of genes is not matched by our knowledge of the gene-environment interaction or potential impacts of biotechnology on the environment. However, many of the Green Revolution technologies were also introduced without such understanding. At present, there is widespread distrust of biotechnology and the public needs to be engaged in dialogue before the technology is disseminated widely.

WATER MANAGEMENT IN RAINFED AREAS

Water management is the activity of planning, developing, distributing and managing the optimum use of water resources. In an ideal world, water management planning has regard to all the competing demands for water and seeks to allocate water on an equitable basis to satisfy all uses and demands. This is rarely possible in practice. About 60 percent of India's cultivable area will remain dependent on dry land farming even after all irrigation potential is fully exploited. Productivity growth in these areas is obviously critical for rural income growth and poverty alleviation, and it depends critically upon better moisture conservation and the development of varieties suited to dealing with moisture stress.

Schemes for water retention, moisture conservation and groundwater recharge have been implemented for many years in India but with mixed results. Experience suggests some pointers for the future. Greater use of technology inputs can help a great deal.

Satellite mapping by the Indian Space Research Organisation (ISRO) has been particularly helpful in planning watershed management schemes to achieve optimal results. It is also important to adopt a holistic approach. For example, if deforestation problems upstream are not tackled, water retention structures downstream will quickly silt up.

Community participation is critical to impart ownership and ensure an acceptable distributional outcome. In the past, these multiple factors were not effectively integrated into watershed development schemes. Now a National Rainfed Area Authority has been proposed to help coordinate the work of different implementing agencies.

The cost of treating rainfed areas to ensure optimum use of available water is approximately Rs.10,000 per hectare and the untreated area is about 80 million hectares, yielding a total cost of approximately US$20 billion. If this amount is added to the cost of irrigation development and the target to be achieved over a 10-year period it would require a doubling of public investment in irrigation.

Other Inputs

Increasing agricultural productivity also depends on the efficient delivery of several other inputs. The quality of seeds and planting material needs to be greatly improved, and this calls for strengthening the research effort to make it more effective. Two expert committees have recently reported on how to restructure the agricultural research system to make it more results oriented and their recommendations are under consideration. The system for producing and marketing certified seeds of existing varieties at reasonable prices also needs to be improved. Seed replacement rates in most parts of the country are only one-third to one-half of what they should be, a situation which reflects partly a lack of knowledge of the importance of seed replacement and partly also a lack of availability of reliable seeds.

There is evidence that the use of fertilizers is at present highly imbalanced suggesting that scientific application of fertilizers holds potential for raising productivity. Nitrogen fertilizers are oversubsidized compared with phosphorus and potassium fertilizer. The structure of fertilizer subsidies should be rationalized to avoid excessive and wasteful use of nitrogen fertilizers. Lack of knowledge of micronutrient deficiency in the soil is also a serious problem.

There is need for much more extensive soil testing to encourage balanced application of nutrients needs. Underlying these problems is the deterioration of the extension services, which makes it difficult to disseminate best farming practices. Strengthening the extension system therefore needs special attention. The government has also identified credit to farmers as a critical area for corrective action. The public sector commercial banks are being pushed to provide credit to agriculture and have responded commendably. However, the cooperative credit system which was meant to be the backbone of agricultural credit, has become financially weak. Part of the problem has been the politicization of cooperative institutions as a consequence of interference by state governments. The central government is considering ways of reviving the cooperative credit system by recapitalizing the cooperative banks provided state governments agree to changes in the system of governance that would ensure professional management of cooperative banks without state government interference.

AGRICULTURAL DIVERSIFICATION

India's future agricultural strategy must also be oriented to the need for agricultural diversification. India's foodgrain production capacity has increased significantly over the years and there is evidence that household consumption patterns are changing away from foodgrain towards higher value crops such as vegetables, fruits, milk, eggs, etc. Future growth in agriculture must come from diversification into these non-food grain areas and this will pose a special challenge because marketing these perishable products is much more complicated than marketing food grains.

Horticulture development is currently constrained by poor marketing arrangements. The gap between prices received by the farmers and those paid by urban consumers is large, reflecting inefficient marketing arrangements. Horticultural produce is typically collected from farmers by market agents, who sell it organised markets established under the Agricultural Produce

Marketing Acts. Unfortunately these markets are controlled by a few traders and operate on highly non transparent basis. Facilities for grading and handling are poor and methods of price discovery in the markets are nontransparent. Wastage is high owing to poor logistics and the absence of cold chains. The net result is much lower realisation by the farmer.

Figure: Agricultural Diversification

It is necessary to amend outdated laws restricting the establishment of markets to allow cooperatives and private entrepreneurs to set up modern markets with grading facilities, cold storages and transparent auction procedures. Half a dozen states have already amended their existing laws on agricultural marketing to allow such markets to be established and a dozen others are in the process of doing so. These changes are being resisted by those who control the existing structure but this opposition will weaken over time.

Contract farming is another innovation that has been introduced in many states and could accelerate diversification. India's laws on agricultural land do not allow corporate bodies to purchase land and operate large-scale farms – a national policy to prevent displacement of a large number of small farmers. – but corporate buyers who know what is needed in export markets, in high end domestic markets or in agro-processing, can engage in contract farming to procure high quality produce. Buyers select areas suitable for the crops they are interested in and organize farmers to produce these crops under contract, while providing planting material of the right quality and technical supervisory. The process enables the farmer to eliminate marketing risk while allowing the corporate buyer to ensure quality supplies by selecting planting material, and provide access to scientific advice on disease and other types of stress.

The development of agro-processing will be a spur agricultural diversification and the government is paying special attention to this area. At present, the proportion of India's agricultural output that is processed is very small compared with that in most developing countries and the demand for processed food is bound to increase as incomes rise. There are several obstacles to the more rapid development of food processing. Taxation structures often discriminate against food processing because processed food is the first stage for application of indirect taxes and the absence of tax rebate on taxes paid on inputs means the effective tax on value added is very high. Another impediment is the reservation of certain categories of products for small scale production. The absence of a modern food processing law has meant that this sector is governed by multiple laws, making it difficult to operate effectively. An Integrated Food Processing Law has been introduced in Parliament to replace and the passage expected in the current year will make a qualitative difference to the operating environment.

AGRICULTURAL DIVERSIFICATION ON SMALL FARMS

Agricultural diversification is an important mechanism for economic growth. It depends, however, on there being opportunities for diversification and on farmers' responsiveness to those opportunities. Agricultural diversification can be facilitated by technological breaks-through, by changes in consumer demand or in government policy or in trade arrangements, and by development of irrigation, roads, and other infrastructures. Conversely, it can be impeded by risks in markets and prices and in crop-management practices, by degradation of natural resources, and by conflicting socio-economic requirements-perhaps for employment generation, or for self-sufficiency or foreign-exchange-earning capacity in particular crops or livestock or fishery or forest products.

PRESENT STATUS OF AGRICULTURAL BIOTECHNOLOGY IN ASIA

Selected Country Experiences

Several governments in Asia are committed to the use of modern biotechnology in agriculture. They have devoted significant human and financial resources to this policy over the past two decades. Some illustrations of current activities in selected countries are given below.

The PRC accords high priority to biotechnology to increase food production and improve product quality in an environmentally sustainable manner. The

PRC has moved quickly to adopt new biotechnologies, particularly genetically modified crops. The country is rapidly increasing its expenditure on biotechnology R&D. Over 103 genes have been evaluated for improving traits in 47 plant species. New traits have been introduced and evaluated in field tests on rice, wheat, maize, cotton, tomato, pepper, potato, cucumber, papaya, and tobacco. A variety of traits have been targeted. They include resistance to diseases, pests, and herbicides, and quality improvement. Approximately 50 genetically modified varieties have been approved for environmental release, or small-scale field testing in the PRC. A few new genetically improved varieties have been approved for largescale commercial production. The most widespread are new pest-resistant varieties of cotton that are being widely cultivated by farmers. These were grown commercially by approximately 3 million farmers on approximately 500,000 ha in 2000. Several new products are in the pipeline for potential commercialization (Zhang 2000). India has allocated large public resources toward human resources development and infrastructure in biotechnology. In the early 1980s, the Government of India created a Department of Biotechnology to promote the use of new biotechnologies in industry, medicine, and agriculture. Current R&D efforts in India are directed toward increasing agricultural productivity, bioremediation in the environment, medical and industrial biotechnology, and bioinformatics (Sharma 2000). R&D priorities in agriculture include new regeneration techniques for the rapid multiplication of citrus, coffee, mangrove, vanilla, and cardamom. Cardamom yield has increased 40 percent through the use of tissue culture.

There is substantial private sector participation in biotechnology in India, for example, in the seed sector, the veterinary products sector, and bioinformatics linked with the booming information technology sector (Dhawan 2001). Further, organizations like the M.S. Swaminathan Research Foundation have developed innovative approaches such as the ADB-supported BioVillages program that is fostering the growth of new income-raising technologies. Emphasis is on developing small-scale bioindustries for women (Lakshmi 2001).

Indonesia has placed a high priority on biotechnology over the past 15 years. The Government has designated three National Biotechnology Centers to coordinate R&D in agriculture, medicine, and industrial microbiology. Applications of biotechnology to agriculture are primarily the responsibility of the Agency for Agricultural Research and Development (AARD). A National Committee on Biotechnology advises the minister in developing guidelines for government policy in the promotion of biotechnology. In recent years there has been an extensive training program within Indonesia and abroad to upgrade skills of scientists involved in biotechnological research. In the 1980s, a major World Bank loan of over $100 million financed the creation of three inter-university centers for agricultural, medical, and industrial biotechnology. More recently, a current World Bank loan is financing facilities for agricultural biotechnology within AARD.

Crop improvement efforts using modern biotechnology started in Pakistan in 1985, when a training course was held on recombinant DNA. Work is now concentrated on chickpea, rice, and cotton. Field evaluation is hampered by lack of biosafety regulations. There is some private investment in R&D of agricultural biotechnology. The government controls testing, multiplication, distribution, and biosafety issues for genetically modified crops. Pakistan lacks firm policy and regulations regarding intellectual property rights (IPR) and patents involving biotechnology, and biosafety regulations for GMOs (Zafar 2001).

The Philippines began its modern biotechnology programs in 1980 with the creation of the National Institutes of Molecular Biology and Biotechnology in Los Baños, with a focus on agricultural biotechnology. In 1997, the Agriculture Fisheries Modernization Act recognized biotechnology as a major strategy to increase agricultural productivity. The Act provided a budget for agricultural biotechnology of almost $20 million Present Status of Agricultural Biotechnology in Asia 21 annually for the next 7 years (4 percent of the total R&D budget). In 1998, the government funded these five high-level biotechnology research projects to develop

- (i) new varieties of banana resistant to banana bunchy top virus and papaya resistant to ringspot virus,
- (ii) delayed ripening papaya and mango,
- (iii) insect-resistant maize,
- (iv) marker-assisted breeding in coconut, and
- (v) coconut oil with high lauric acid content.

Nongovernment organizations (NGOs) and other groups concerned about the safety of GMOs have been vocal in the Philippines. This is affecting field-testing and commercialization of transgenic crops. Products in the regulatory pipeline include new varieties of insect-resistant maize and insect-and disease-resistant rice. Thailand is focusing on the applications of biotechnology to traditional foods, fruits, and export commodities such as shrimp. R&D priorities are to increase production and reduce production cost on crops such as rice, cassava, sugarcane, rubber, durian, and orchids. An early success in Thailand has been the development of new molecular diagnostics for the diagnosis and control of virus diseases in shrimp. These diseases cost the shrimp export industry over $500 million in lost production in 1996. The development and commercial use of the new diagnostics prevents the loss of an estimated 20–50 percent of annual production, a saving of at least $100 million per year. There are also active agricultural biotechnology programs in Bangladesh, Malaysia, Nepal, Sri Lanka, Singapore, and Viet Nam.

Regional and International Programs

International Agricultural Research Centers

International R&D programs using modern biotechnology are being conducted by the international agricultural research centers (IARCs), particularly the International Rice Research Institute (IRRI), International Maize and Wheat Improvement Center (CIMMYT), International Crops Research Institute for the Semi-Arid Tropics (ICRISAT), International Livestock Research Institute (ILRI), and International Service for National Agricultural Research (ISNAR). The Center for International Forestry Research also uses biotechnology in the characterization of forest diversity in its Asian program. The International Center for Living Aquatic Resources Management is using new technologies in the improvement of fisheries and aquaculture systems. ILRI is initiating a program on Asian livestock improvement. The CGIAR centers invest approximately $30 million per year in modern biotechnology.

ISNAR Biotechnology Service

The ISNAR Biotechnology Service (IBS), with Japanese support, has been assisting selected Asian countries in developing human resources for managing biotechnology research programs or institutions. IBS has developed specialized courses to enhance the capacity and competency of managers, focusing on strategy building, priority setting, managing biosafety and regulatory aspects, resource generation and deployment, product delivery, and information sharing as well as the establishment and management of linkages.

Acquisition of Agri-biotech Applications

The International Service for the Acquisition of Agri-biotech Applications (ISAAA) is brokering public-private sector partnerships to facilitate technology transfer. It has current regional projects to increase the productivity of maize, papaya, and sweetpotato. ISAAA is also establishing a new Asian knowledge center for crop biotechnology, based in the Philippines, in partnership with CAB International and the Southeast Asian Regional Center for Graduate

Study and Research in Agriculture. Its purpose is to make available timely and balanced information on the risks and benefits of crop biotechnology to interested parties in Asia. In doing so, it aims to provide training and study tours highlighting experiences not only in research but also with biosafety and intellectual property issues. Present Status of Agricultural Biotechnology in Asia

Activities of Funding Agencies

The major external sources of assistance for agricultural biotechnology in the Asia/Pacific region are ADB, Australia, the Rockefeller Foundation, the United Nations, the United States Agency for International Development (USAID), and the World Bank.

Asian Development Bank

ADB has made several strategic and innovative investments in agricultural biotechnology over the past decade. These investments have been in the form of loans and technical assistance projects.

Agriculture and Science and Technology Loan Projects

Several governments have requested ADB financial support for human resource development, laboratory facilities, and equipment for agricultural biotechnology programs. These programs have been integrating new applications of biotechnology into their conventional agricultural R&D programs.

The applications include the use of new molecular diagnostics for pests and diseases and MAS for crop and livestock breeding. Such components are being supported under ongoing ADB projects in the Philippines, Sri Lanka, and Thailand. ADB has provided other loans and grants in closely related areas such as in integrated pest management in cotton under the Cotton Development Project in Pakistan.

ADB Regional Technical Assistance Projects

ADB has provided regional technical assistance grants for the development of the three regional crop biotechnology networks over the past decade. The networks are the Asian Rice Biotechnology Network (ARBN), initiated in 1993, the Asian Maize Biotechnology Network (AMBIONET), initiated in 1998 and the recently established Asian Semi-Arid Crops Network. These networks have been influential in developing capacity to use new techniques in crop breeding for the major cereal crops (rice, maize) and the crops important in the semi-arid regions (sorghum, pigeonpea, and groundnut). The networks are managed by three IARCs (IRRI, CIMMYT, and ICRISAT) that work with national re search institutes to implement the programs on rice, maize, and semi-arid crops.

A recent evaluation of ADB's research investments through the international centers reported favorably on the achievements of the rice and maize biotechnology networks, and recommended further support for this type of research cooperation. The networks provide a platform by which countries and the IARCs can collaborate in the use of the new tools of functional genomics to identify genes that control important traits such as drought and salinity tolerance. This will speed up the breeding of crop varieties with these characteristics that have been difficult to address through conventional breeding. There will be increasing opportunities for cooperation between the networks, as more knowledge is gained on the commonality of genes between species and their control.

Australian Government

Over the last 15 years, ACIAR has entered into more than 100 biotechnology R&D partnerships with at least 10 Asian countries in support of more than 600 active or completed projects. The emphasis of this work has been on developing diagnostics and vaccines for a large suite of diseases of tropical livestock, with some recent work on fish and shrimp being undertaken. Most of these projects have been implemented through government programs, but NGOs are becoming increasingly involved. Molecular marker methods for identifying disease-resistant genes and prolificacy in livestock have also been developed. Several projects on biotechnology for rumen manipulation have been carried out.

Crops and forestry work has focused on development of diagnostics for diseases (viral, fungal, mycoplasma, and bacterial) and contaminants in tropical crops, and the application of biofertilizers, bioremediation technology, and biofumigants. Molecular markers have been developed for the improvement of cereals and tree species. In cooperation with IRRI, attempts are being made to develop apomixis systems for rice. Tissue culture methods for micropropagation and conservation of several species, including sweetpotato, taro, tropical fruits, coconut, green tea, and tree species such as mangrove are being developed.

Eight of ACIAR's current or completed projects have included aspects of plant genetic engineering, with target crops being cereals and pulses, groundnut, and several tropical fruits. Target characters include virus resistance and quality defects related to ripening processes. These collaborative projects were developed at the request of the Asian countries, which fully approved regulatory procedures. ACIAR also provides core funding to many CGIAR and other IARCs, a proportion of which is applied to biotechnology R&D.

Australian support has been provided through AusAID for an Association of Southeast Asian Nations (ASEAN)/Australia regional biotechnology network, mainly concerned with food, microbial, and industrial biotechnology. AusAID also supports the characterization and conservation of genetic resources of taro and forest genetic resources in the South Pacific. AusAID has supported several biotechnology seminars. A major seminar in 1989 at the Australian Academy of Sciences on the potential of agricultural biotechnology in international development reported on the outcomes of a joint study cosponsored by AusAID, ACIAR, ISNAR, and the World Bank (Persley 1990a, 1990b).

In the area of human resources, the Crawford Fund has sponsored several master classes in biotechnology for senior policymakers over the past decade.

Agency for International Development

The USAID is supporting applications of biotechnology through bilateral activities in several countries, including India, Indonesia, and Pakistan. USAID is also providing specific support for biotechnology applications within the programs of IARCs. This includes support for research by IRRI on improving the nutritional quality of rice by increasing its vitamin A and iron content, and support for research by the Tata Energy Research Institute on the development of *golden* mustard.

World Bank

The World Bank has supported the development of infrastructure and human resource development for biotechnology in several Asian countries over the past 15 years. This support has come through loans in the agricultural sector, science and technology, and education. There are currently substantial components for biotechnology within agricultural technology projects in India, Indonesia, and Pakistan. The Inter University Centers for Biotechnology in Indonesia were built with a $150 million loan in the 1980s. The Republic of Korea also used World Bank loans to develop its infrastructure in biotechnology. The World Bank is also one of the main financial supporters of CGIAR centers. The centers invest about 10 percent of their total annual budget of $340 million in the applications of biotechnology.

The Rockefeller Foundation

Operational from 1984-1999, the Rice Biotechnology Network sponsored by the Rockefeller Foundation was very active and successful in the region. The program's two objectives were to (i) to create biotechnology applications to produce improved rice varieties suited to developing country needs, and (ii) to train scientists in developing countries to use the techniques and adopt them to their own objectives. A network of about 200 senior scientists and 300 trainee scientists from all the major rice producing countries of Asia and a number of industrialized countries participated. The program was funded at approximately $5 million per year for 15 years.

Another outcome of the program was the development of promising new technologies for the control of rice pests and diseases and improving the nutritional quality of rice. The work is being continued by national governments, ARBN, and IRRI to do the field evaluation and distribute new rice varieties in Asia. The Rockefeller Foundation presently concentrates its biotechnology programs in Asia on drought tolerance in rice and maize. It is also examining innovative means of dealing with access to technologies and IP issues.

United Nations Agencies

The Food and Agriculture Organization (FAO) is giving high priority to biotechnology within its Asia/Pacific regional programs (FAO 2000). During 1989-1993, the United Nations Development Programme financed the establishment of biotechnology centers in eight countries (PRC, India, Indonesia, Republic of Korea, Malaysia, Pakistan, Philippines, and Thailand) to share rDNA techniques in animal improvement, embryo transfer, and disease control. In 1994 the United Nations Industrial Development Organization (UNIDO) established an International Center for Genetic Engineering and Biotechnology in New Delhi, India, to assist Asian countries in the applications of biotechnology to important crops of the region.

RISK ASSESSMENT AND MANAGEMENT

Public Perception of Risks

As with any science and technology, biotechnology can bring with it benefits and risks. It is the risks of agricultural biotechnology that have received widespread publicity in the media even though biotechnology has also been applied to health and industrial sectors. Environmental NGOs have been particularly vocal in taking issue with the new technologies derived from or incorporating GMOs. As a consequence, in the public debate biotechnology has become synonymous with GMOs, although they are only one of the many products of biotechnology.

Curiously, biotechnology and GMOs in health-care products now in widespread use (insulin, hepatitis vaccine, medication for cardiovascular disease, etc.) or for industrial purposes such as bioremediation have elicited no such controversy.

This can probably be attributed to the lack of benefits to consumers in the first generation of genetically modified (GM) crops. The main focus was on herbicide and insect resistance that benefited farmers, seed producers, and chemical companies. It is expected that the next generation of genetically modified foods will benefit consumers, nutritionally or from taste or storage benefits, and accordingly may be better accepted.

A number of food-related crises in recent years have made consumers particularly sensitive about food safety issues.

Health and food safety concerns are again at the forefront in Europe following additional cases of mad cow disease *(bovine spongiform encephalopathy)* and the banning throughout the European Union of blood and bone meal in feed for all animals.

These crises have not been caused by GMOs, but by the intensification of agriculture and food production, a fact that appears to have escaped public attention. In Europe in particular, demands have been made for informative food labelling so that consumers may, if they wish, avoid genetically modified foods.

The anti-GMO movement reveals profound mistrust of developments in science and technology and of the forces seen to be driving them. Genetically modified crops such as maize, sorghum, cotton, and canola have been widely grown for the last five years, yet no harmful effects on human health or the environment have been detected. That was one of the conclusions of the OECD-sponsored conference held in Edinburg, UK, in 2000. However, it is generally agreed that government and the private sector are responsible for monitoring the long-term effects of GMOs on human health and the environment.

Potential Risks of Biotechnology

The risks associated with modern biotechnology fall into four categories: food safety, environmental, socioeconomic, and ethical. Some of these concerns relate to potential risks inherent in modern biotechnology and can be described as technology-inherent (Leisinger 2000). Others are related more to value systems or cultural practices and can be described as technology-transcending.

Food Safety Concerns

The potential risks of biotechnology on human health may include toxic reactions, increased cancer risks, food allergies, food contamination, and antibiotic resistance.

There is also concern that GMOs in animal feed might present a health risk for consumers, or for the animal itself. Consumers are also concerned about the long-term health effects of genetically modified foods.

To address food safety concerns, the following safeguards have been adopted by some countries:

(i) Some countries have regulatory procedures, institutions, and infrastructure in place to ensure food safety (OECD 2000). These regulations cover all aspects of the food chain, from farm inputs (including animal feed, feed additives, pesticides, fertilizers, veterinary drugs) through production and processing (including agricultural products, processed food, novel foods, food additives), to transportation, storage, and distribution.

(ii) Formal science-based procedures for risk analysis of food have been adopted by some countries. These continue to evolve with new scientific information about food safety, emerging pathogens, new technology, and consumer demands for a high level of public health protection. Generally, the procedures conform to international standards set by the Codex Alimentarius Commission of FAO.

(iii) Some countries have adopted a variety of approaches to regulate genetically modified foods, either by applying existing food safety measures or enacting new legislation.

(iv) In response to consumer demands, a growing number of countries are introducing labelling. Opinions differ on whether labelling should be mandatory or voluntary, as well as on acceptable tolerance levels and on the type of information to be used. Two labelling issues currently being addressed are segregation and traceability. Segregation relates to the ability to attest to the separation of genetically modified and non-genetically modified crops (as in the case of soybean in processed foods). Traceability means being able to attest to the origins of food products. Traceability is of particular importance in the current controversy related to mad cow disease. Low cost, quick tests are now available to determine the presence of genetically modified products in food.

Environmental Concerns

A number of biotechnology applications are not seen to present any new threats to the environment. That is the case with tissue culture, diagnostics, and market-selected plant breeding. On the other hand, there is fear of potential risks from the release of GMOs into the environment. The potential risks of GMOs on the environment may include increased pesticide residues, genetic pollution, damage to beneficial insects, creation of superweeds and superpests, creation of new viruses and bacteria, and genetic bioinvasion.

To address environmental concerns, some countries have adopted the following safeguards:

(i) Most current biotechnology applications in tissue culture and micropropagation, diagnostics and vaccines, and markerassisted plant breeding are subject to existing regulations. They include phytosanitary regulations and plant quarantine, varietal certification of seeds, and veterinary product regulations. These regulations usually conform to international standards, guidelines, or recommendations such as those set by the International Plant Protection Convention or the International Bureau of Epizootics for animal products.

(ii) A number of countries have introduced new requirements and procedures for the environmental release of GMOs. Procedures for hazard identification and risk assessment of GMOs are now well-established in most OECD countries, and in some Asian countries.

Socioeconomic Concerns

Modern biotechnology R&D has been conducted in an institutional and economic environment that differs significantly from the development of the earlier Green Revolution technologies. While the latter were essentially the prerogative of public research institutions and philanthropic foundations, developments in biotechnology have been driven essentially as a competitive, commercial endeavor in which powerful private sector actors compete.

The major socioeconomic risk of agricultural biotechnology stems from the fact that the research, development, commercialization, and distribution of new biotechnological products have been carried out mainly in developed countries by a few, large, multinational companies. These companies have focused on temperate crops for large farmers in developed countries. Undertaking R&D on Asia's basic food crops for small farmers in rainfed and marginal areas is of little interest because they see limited returns from such investments. If this trend continues, modern biotechnology will aggravate the income disparity between developed and developing countries, and between large and small farmers. Increased public investments in agricultural biotechnology are necessary to ensure that small, poor farmers have access to biotechnology. Governments must also address the potential gender disparity in the access to technology, and the negative impact specific to women, since no technology is gender neutral. Lessons from earlier agricultural technological changes should be used as a caveat.

Ethical Concerns

Ethical issues may stem from the uneven or inequitable impact of an expanding global economy; from the national social, economic, and politiRisks Assessment and Management 41 cal context; or from individual values. Indeed, it may be difficult to reconcile personal values with what a majority regard as the common good. One of the ethical concerns raised by biotechnology—and particularly GMOs—is that it is unnatural and an unwarranted tampering with nature. However, seen in historical perspective, most technology developments in agriculture over the centuries have involved, in one way or another, efforts to overcome the vagaries of nature.

Risk Management

To address the potential risks of biotechnology in agriculture, there is a need for Asian countries to establish effective biosafety procedures for the development, testing, and release of the technologies.

Emerging Regulatory Systems for Biotechnology

A growing number of international organizations have become involved in food and environmental safety. Concerted international efforts to develop agreed upon scientific approaches to biosafety date from 1975, when a group of scientists from the United States National Academy of Sciences expressed concern regarding the potential biological risks of rDNA molecules. Efforts to ensure the science-based, case-by-case risk identification, assessment, and management of GMOs have thus been pursued for some 25 years.

In the industrialized world, OECD has been instrumental in forging agreements on scientific principles regarding the safe applications of rDNA organisms in industry, agriculture, and the environment. Since the 1980s, OECD has established guidelines, first for (i) laboratory-based experimentation, then (ii) for small-and large-scale field trials of genetically modified plants and organisms, and finally (iii) for their commercialization and release into the environment or food chain. During the 1990s, a growing number of international organizations including the United Nations Environmental

Programme (UNEP), UNIDO, World Health Organization (WHO), and FAO have become involved in activities related to biosafety. Both UNIDO and UNEP have developed biosafety guidelines. FAO and the WHO are jointly responsible for the Codex Alimentarius Commission, which sets international standards for food safety. Codex has set up a new Intergovernmental Task Force on Foods Derived from Biotechnology to develop general principles for risk analysis and to provide guidance on risk assessment.

Other international organizations conducting work on the safe use of biotechnology include the International Centre for Genetic Engineering and Biotechnology, the United Nations Conference on Trade and Development, and, for many aspects of animal as well as human health, the International Bureau of Epizootics.

Cartagena Biosafety Protocol

Preoccupation with the protection of biological diversity, necessary for sustaining agriculture and food production and, indeed, life itself, became a topic of international debate during the 1980s. As a result, FAO established the International Undertaking on Plant Genetic Resources as a nonlegally-binding agreement for cooperation in the conservation of genetic material. The agreement was based on the universally accepted principle that plant genetic resources are a heritage of mankind and consequently should be available without restriction. The legally binding Convention on Biological Diversity, which came into force in 1993, encompasses not only plant genetic resources but all living organisms. In contrast to the FAO undertaking, the Convention affirms that States have sovereign rights over their own biological resources.

In January 2000, the legally binding Cartagena Protocol on Biosafety was adopted. This Protocol lays the foundation for a global system for assessing the impact of GMOs on biodiversity. It outlines the obligations of all countries that are signatories to protect biological diversity from the potential risks imposed by living modified organisms resulting from modern biotechnology. At the heart of the Protocol is the concept of advanced informed agreement to the import of GMOs. This means that an exporting country must inform an importing country of its intention to export. It must also provide information on the GMOs, including an appropriate risk analysis. For its part, an importing country is required to give its informed agreement to the import. This implies that, although signatory states have the right to prohibit the import of GMOs if advanced informed agreement has not been given, they must also be capable of determining the validity of a safety assessment.

In addition, the Protocol advocates the "precautionary approach," described in the 1992 Rio Declaration on Environment and Development as a principle of international environmental law. The Protocol also proposes to establish a Biosafety Clearing-House for collecting, sharing, and disseminating information on risk assessment and management, which is particularly important for developing countries. Thus far, 81 countries have signed the Protocol, including Bangladesh, PRC, Indonesia, Republic of Korea, Malaysia, Philippines, and Sri Lanka.

Approaches in Selected OECD Countries

A sample of industrialized countries shows a diversity of approaches to food and environmental safety and in the agencies or ministries that play key roles in the national biosafety system. For example, the US has chosen to develop biosafety capacity within existing institutions (notably the Environmental Protection Agency and the Department of Agriculture). Denmark has designated the National Forest and Nature Agency as the lead agency for biosafety. In Norway, in accordance with the Gene Technology Act, the application of GMOs is subject to ethical and social conditions. A GMO can be approved for use only after the assessment of its benefit to the community and its contribution to sustainable development. In Australia, apart from the different ministries involved, an Advisory Committee on Genetic Engineering was created some years ago.

That will soon be replaced by an office with an enforcement rather than an advisory role. A public hearing must be held before authorization is given to release a GMO. Australia has recently created the Office of the Gene Technology Regulator to operate in the Commonwealth Department of Health and Aged Care. Japan has based its risk assessment system on the principles and concepts developed by OECD. Different Japanese ministries and agencies have developed guidelines in accordance with their responsibilities on a product-sector basis. A growing number of countries are introducing guidelines or mandatory measures for labelling genetically modified foods. In the US, formerly strongly opposed to labelling, the Food and Drug Administration has recently been charged with developing guidelines for voluntary labelling of food products to indicate whether or not they contain GMOs. In the European Union, food products containing more than 1 percent GMO must be labelled, stating clearly: "this product contains GMOs." The Australia New Zealand Food Standards Council is drawing up a new Food Standards Code. Labelling requirements will be drawn up in accordance with this joint Code. Japan is planning the introduction of mandatory food labelling in April 2001.

The OECD has also played an active role in promoting the harmonization of biosafety procedures among its member countries so that information and data gathered in the course of risk assessment may be mutually acceptable.

Emerging Regulatory Systems in Asia

A growing number have set up national biosafety systems; others are formulating national biosafety guidelines. Most countries have not yet completed guidelines for commercialization. The PRC is the only country in Asia where transgenic crops have been approved and released for sale. A number of countries are also in the process of introducing food labelling. Indonesia is planning to introduce mandatory labelling in April 2001.

ASEAN has undertaken a number of initiatives aimed at the harmonization of biosafety procedures among its member countries. In 2000 it also conducted a joint workshop with the Ministry of Science and Technology of the PRC to discuss issues related to transgenic plants. The Asia-Pacific Economic Cooperation has a number of capacitybuilding and technical cooperation activities in support of risk assessment, risk management, risk communication, and public acceptance. Responsibility rests with its Agricultural Technical Cooperation Experts Group and its Sub-Group on Research, Development and Extension of Agricultural Biotechnology.

Although some Asian countries have regulations in place, weak law enforcement sometimes results in the release of GMOs without proper procedures having been followed. For example, P.T. Monagro Kimia (a subsidiary of Monsanto) released Bt cotton in Indonesia without complying with Indonesian environmental law on safety procedures, environmental assessment, and information disclosure. Law enforcement in Asia needs strengthening to avoid risk to human and environmental health.

Biosafety Capacity Building in Asia

A growing number of initiatives are being taken to provide support for Asian countries in developing their capacities to safely manage biotechnology. A representative sample follows.

Following the signing of the Biosafety Protocol, UNEP, through its Global Environment Fund, is preparing a major capacity-building program for developing countries. It will also be responsible for the establishment of the Biosafety Clearing-House. FAO serves as the secretariat for the International Plant Protection Convention, which has recently established a working group on the Phytosanitary Aspects of GMOs, Biosafety and Invasive Species. UNIDO has conducted training courses in biosafety in a number of Asian countries.

In collaboration with OECD, it has compiled an information source and database (BIOBIN), which provides information on biosafety methods and procedures, and on the status of the regulatory frameworks in individual countries. OECD has established a network of international organizations with activities related to biosafety to enhance the exchange of information and to facilitate cooperation.

IBS provides a number of services to enhance developing country capacities to manage agricultural biotechnology. It gives short training courses annually for managers of agricultural biotechnology research programs in Asia. Biosafety is a key element in these courses. IBS also provides advice to individual countries.

USAID is involved in a number of biosafety training activities in Asian countries, some of which have been provided by its Agricultural Biotechnology Services Program. ISAAA is currently setting up a Global Knowledge Center on Crop Biotechnology based at ISAAA's SEAsia Center at Los Baños in the Philippines. ISAAA also organizes capacity-building workshops and training for regulatory officials and scientists from the Asian region.

FUTURE CHALLENGES

The population of Asia is projected to increase from 3.0 billion to 4.5 billion in the next 25 years. During the same period, the urban population will nearly double from 1.2 billion to 2.0 billion, as rural people move to the cities in search of employment. These increases will place massive pressure on developing member countries (DMCs) of ADB to increase food production. Food demand is influenced by population growth, urbanization, income, and associated changes in dietary preferences. Urbanization and income growth frequently lead to shifts from a diet based on root crops (cassava, yam, and sweetpotato), sorghum, millets, and maize to rice and wheat, which require less preparation time, and to more meat, milk, fruits, vegetables, and processed foods. This dietary transition has already happened in much of the region (ADB 2000b). Meeting the food needs of Asia's growing and increasingly urbanized population requires increases in agricultural productivity and matching these increases to dietary changes and rising incomes.

To meet this demand, cereal production will need to be increased by at least 40 percent from the present level of about 650 million tons annually, most of which will have to come from yield increases. In addition, meat demand will double during the period (Pinstrup-Andersen et al. 1999). Production increases will have to be achieved by increasing yields in a sustainable way to conserve diminishing and degraded natural resources. Nearly all of these production increases will need to take place in DMCs themselves because on average 90 percent of the world's food is consumed in the country where it is produced. Food imports are not only expensive but discourage the creation of employment, which is badly needed in the rural areas.

In this millennium, we face a food, feed, and fiber production challenge in highly complex farming systems for several reasons:

(i) Water will become the most important limiting factor in agricultural production because the quality and quantity of water will decline as a result of pollution, forest degradation, and increased agricultural, domestic, and industrial use (ADB 2001).

(ii) Urbanization will mean the loss of agricultural land to residential and industrial development, and a decline in the number of farm workers.

(iii) Most farmers are poor with small landholdings.

(iv) Farming systems are commonly heterogeneous with mixes of food crops, livestock, and trees.

(v) About 70 percent of the cultivated land is rainfed with unreliable distribution and intensity of rainfall.

Thus, the increase in food production during the next 25 years will have to be achieved using less labour, water, and cultivated land. This can be done only if scientists can develop new crop varieties with high yield potential and high water use efficiency. New understanding of plant and animal genes may offer ways to increase crop yields to the levels required to adequately and sustainably feed the growing population in Asia. Thus, developments in modern biotechnology could make extremely important contributions to future agricultural growth, food security, and poverty reduction. Increasing smallholder agriculture productivity will not only increase food supplies, but will reduce poverty and malnutrition, increase food access, and, improve living standards of the poor (McCalla and Brown 2000).

2

Agriculture Biotechnology in the Changing Global Scenario

INTRODUCTION

Steady globalization of trade has profound implications for future agricultural development. The diversity of India's agro-ecological setting, high bio-diversity and relatively low cost of labour provide potential for agricultural competitiveness in a globalized economy. It is expected that with increasing globalization of markets over the years there will be demands for agricultural intensification. This will also be favoured because of greater backward and forward linkages between agriculture and food industry.

Therefore, increase in production and productivity are bound to be strategically important to economy. Intensification will not only favour alleviation of rural poverty but will also improve resource conservation particularly in the small farming sector where farmers can be encouraged to take up organized production of high value crops such as fruits, specialty vegetables, flowers medicinal and aromatic herbs etc.

Stronger demands for crops of the small farmers' will not only improve incomes and welfare but will also make investments in technology and resource conservation more attractive. The General Agreement on Tariff and Trade (GATT) and liberalization of global trade is bound to have impact on future land use and production pattern. Understanding the local, national and international environment under which agricultural production is taking shape will be crucial in developing our own strategies.

Extension Strategies

Since early fifties a number of public by funded agricultural development programmes have been sponsored. These have included programmes like the

National Extension Service (NES) Blocks in 1953, the Intensive Agricultural District Programme (IADP) in 1961-62, the Intensive Agricultural Area Programme (IAAP) 1964-65, the High Yielding Variety (HYV) programme 1966-67 and the Small and Marginal Farmers' Development Programmes (SMFDP) in 1969-70.

Though these programmes had a perceptible impact the efforts did not get replicated over different areas and categories of farmers. In mid seventies based on pilot level project in Rajasthan Canal and Chambal command area a 'Training and Visit' (T&V) system of extension was promoted in different states. Extension efforts of the Indian Council of Agricultural Research through its research Institutes and the State Agricultural University were largely limited to demonstration of new technologies through such programmes as National Demonstration Project, Operational Research Project, the Lab to Land Programme and the Krishi Vigyan Kendras.

However, there appears much to be desired in the way that extension programmes are conceived and implemented. At present extension programmes are implemented in largely a top-down fashion leaving little scope for localized planning and action.

Farmers are almost passive receivers and their involvement in the process of technology generation and adoption is almost absent. Extension services, at present, are almost exclusively in the public sector domain and there is no effort or institutional support for other operators *e.g.* the NGOs, the corporate bodies etc.

Extension programmes sponsored by the government operate largely in isolation and there appears a strong need to view the extension programmes as an integral part of the research and development process. The challenges facing agricultural development call for fundamental changes in our approach to technology transfer/extension programmes. Changes are necessary in the context of changing economic environment following policy adjustments in relation to privatization, deregulation and globalization calling for greater efficiency and effectiveness of the extension system.

More importantly there is need for:

- Greater emphasis on providing producers with knowledge and understanding needed to overcome the problems or to exploit opportunities of their own specific production systems. Correspondingly there will be a need to de-emphasize 'package of practices' or the blanket recommendations, top down approach followed thus far.

- Shift in the focus of public extension systems from promoting inputs use to one on sustainable management of resources and improvements in the production system as a whole.
- Closer interaction between farmers, extension scientists and production system researchers in diagnosing problems and identifying location specific recommendations emphasizing participation and education rather than being prescriptive.
- Widening the range of extension delivering agencies. While the publicly operated extension systems will continue to be important, there will appear a greater role for NGOs, farmers' associations and corporate sectors in particular situations. Role of commercial suppliers of seeds, agrochemicals, machinery, vaccines and medicines in providing advisories, as is already being done in a limited way, will need to be encouraged and factored into public system's own priorities.
- Wider and more creative use of mass media in tune with current developments in information technology to get information across to the farming community whose ability to overcome constrains at farm level will increasingly depend on access to reliable and up-to-date information.

BIOTECHNOLOGY FOR SOCIETAL DEVELOPMENT IN INDIA

The application of biotechnology for societal development has always received special attention since it benefits socially and economically disadvantaged sections of the society. Through this programme a number of projects are implemented for economic and technical empowerment of the beneficiaries. Efforts are made through training, demonstration and appropriate R&D inputs for providing tailor made solutions for location specific problems of SC/ST, women and rural population using modern biology and biotechnology. The following pages provide the highlights of the achievements.

Programmes for SC and ST Population

Biotechnology-based Programme for SC/ST population has benefited around 15,000 people directly and about 45,000 indirectly through 54 projects during the year. The projects have been supported in the areas of cultivation of medicinal and aromatic plants, biofuels, production of biofertilizers and biopesticides, organic farming, aquaculture, mushroom cultivation as well as human healthcare interventions etc. The universities, public funded institutions, Krishi Vigyan Kendras and voluntary/non-governmental organizations have been involved in the implementation of the projects. The target groups have received the benefit of hands-on training and field demonstrations.

Vermicomposting: MR Morarka GDC Rural Research Foundation Jaipur implemented a project in Jalgaon district of Maharashtra, to benefit 2600 SC/ST target population on economic utilization of waste, agriculture residues and cow dung. Over 1780 MT of waste, agricultural residues and cow dung was utilized for vermicomposting through establishment of 28 community facility centres.

Beneficiaries are utilizing vermicompost on crops like onion, sugarcane, cotton, banana, papaya and wheat. Some farmers switched over from conventional crops to cash crops such as colorful capsicum, gerbera as well as inter cropping of safed musli, turmeric and ginger in banana plantation.

Farmers have been trained to use bio-fertilizer, foliar sprays and mulching to use the agro-waste for fertility and management. Marketing linkages have been established with "Organic Shoppe", through which farmers are getting 10-20% premium for their organic produce. About 20 entrepreneurs have taken up vermiculture as commercial activity. extension Material in the local language was distributed to the farmers.

Mushroom cultivation: A project on promotion & production of essential oil, mushroom and vermicompost was implemented by RRL, Jorhat for revenue generation of the ST population of Arunachal Pradesh. Motivation and demonstrationcum-training programmes have been conducted in 13 villages of five districts on cultivation of citronella, production of mushroom and vermicompost.

Demonstration units have been established at RRL Branch Naharlagun farm. Mushroom spawn, earthworms and citronella saplings were distributed to the beneficiaries. Several beneficiaries established vermicompost units on commercial basis and are selling vermicompost. The beneficiaries at Ziro, Jairampur, Doimukh, Manmao, Hetlong and Tengman have established oyster mushroom cultivation units. So far, 10 acres of land have been planted with citronella and 7 acres of land with patchouli at Sonajuli village of Papum Pare district. Medicinal plants: A project on strengthening health care practices and conservation of medicinal and nutritious plants was continued at Herbal Folklore Research Centre, Tirupati for implementation in 100 villages of two Mandals in Andhra Pradesh.

Training workshops were organized with the help of Yanadi tribal people for community health volunteers on identification and sustainable harvesting methods of 20 species of locally used medicinal and aromatic species. The reputed healers were encouraged to prepare medicines needed for PHC from the identified plants. About 6000 seedlings were distributed to the beneficiaries for planting in kitchen gardens. Green health kits (samples of herbal medicines) and educational material in local language containing identification, cultivation techniques and preparation of medicines with their therapeutic properties were distributed to the trainees.

A project on creating awareness and promotion of medicinal plants for the production and marketing was implemented by RRL, Jorhat to benefit ST population of Arunachal Pradesh. Awareness and training programmes were organized on the importance, scientific cultivation and marketing of Piper longum, Kaempferia galanga, Andrographis paniculata, Stevia rebaudiana, Rauvolfia serpentina and Mentha piperita.

Three nursery-cumdemonstration centers were established at Roing (1000 ft msl), New Bomdila (8160 ft msl) and Ziro (4350 ft msl) and 19000 saplings were distributed to ST farmers/NGOs for introduction and propagation in their fields. At high altitude, villagers were advised to cultivate Aconitum heterophyllym, Coptis teeta and Picrorrhiza kurroa.

Beneficiaries have stopped collecting wild Homalonema aromatica, an important but rare aromatic plant of Arunachal Pradesh, and have started its cultivation in their farms. Ten community nurseries have been established with the help of Vivekananda Trust, Roing at ten villages of Lower Dibang Valley District. Twenty farmers were trained on cultivation and post-harvest management of medicinal plants and marketing tie-up has been established. Around 1266 local ST people participated in the awareness programmes.

A project on organic cultivation of medicinal and aromatic plants for income generation and sustainable development in Uttarkashi District of Uttaranchal was undertaken through Forests Conservation Programme of World Wide Fund for Nature, New Delhi. Demonstration-cum-Resource Material Production Centre established for training on vermicomposting and cultivation of species.

Construction of poly house, vermi-composting units and planting of seeds of Rauvolfia serpentine completed. Awareness and motivation camps were organized for the farmers. Extension material brought out in the local language for cultivation of Rauvolfia serpentina, Coleus barbatus and Pelargonium graveolens were distributed to the farmers.

A project on developing skills and economically sound vermicompost units utilizing distillation waste of Mentha arvensis (menthol mint) was implemented for the benefit of SC/ST communities of Lucknow and Barabanki Districts of Uttar Pradesh. Around 200 farmers were trained on menthol mint nursery raising, transplanting, distillation and practices followed in vermicomposting. Demonstrations were carried out in farmers' field using high yielding variety 'Kushal'. Farmers trained on organic farming started vermicompost production using distillation and other agro-waste. The yield data obtained from the demonstrations showed the possibility of reduction of chemical fertilizers by 75% without affecting the yield.

Biofuels

A project was continued on cultivation of Jatropha in arid lands of Namakkal District in Tamil Nadu for production of biofuel and other products for SC/ST community by Centre for Research in Social Sciences, Coimbatore. A high tech nursery was established and plantlets of Jatropha were raised. Training programmes were organized for 223 members on cultivation of Jatropha in 31 acres of land.

About 30,000 plants were supplied for cultivation. Awareness camps, training and demonstrations were conducted on Jatropha cultivation, oil extraction and soap making. SHG groups were trained in nursery techniques. Women self help groups were trained in various activities of jatropha cultivation and soap making. Extension material bought out in the local language was distributed to the beneficiaries. Biocontrol agents: A project was implemented on development of eco-preneurship among the coastal community of Pondicherry by MS Swaminathan Research Foundation, Chennai. Sixteen SHGs involving women and youths were formed in Ariyankuppam, Nonankuppam, Kakayanthope and Veerampattinam villages. Demonstrations and trainings were conducted to educate trainees on basic aspects of IPM, importance of biocontrol agents, production and marketing of Trichogramma, importance of human nutrition, role of edible mushrooms and production & marketing of oyster mushroom.

Micro-enterprise was developed for mushroom and Trichogramma production. Technologies were disseminated on biopesticide (Trichogramma), oyster mushroom production and vermicomposting to recycle the spent waste of oyster mushroom. Extension material brought out in the local language was distributed to the villagers.

Aquaculture

A project was undertaken on enhancing livelihood of SC/ST people in Orissa and Tamil Nadu by MS Swaminathan Research Foundation, Chennai for introducing various aquaculture enterprises to 250 participants in backyard ornamental fish culture and carp seed production. Fingerlings produced were distributed to the entrepreneurs for stocking. Various aspects on ornamental fish breeding, carp nursery and gold fish culture were disseminated to the self help groups, which helped them in earning a revenue of Rs. 200/month/ household and an income of about Rs. 3200/from carp nursery and Rs. 8500/ from vegetables and flowers as bund crops. Community ponds were effectively utilized by stocking with rohu, catla, mirgali, Chinese & Indian carps and marketable size fish was harvested. Technologies transferred to the entrepreneurs on community pond aquaculture, carp nursery and ornamental fish culture and extension material in local language were distributed to the beneficiaries.

A project on utilization of integrated marine living resources by SC youth from Tuticorin region of Tamil Nadu was undertaken by College of Fisheries, Tuticorin. Lobster fattening and crab culture techniques were standardized and demonstrated to the beneficiaries, who have grown lobsters in cages for about 4 months and produced 250-350gms size for marketing. The beneficiaries were helped in availing credit facilities and bank loan. In the project around 120 people of target population have been trained. A manual on crab and lobster fattening technology was brought out for the distribution to the beneficiaries.

A project on freshwater prawn culture was implemented in village Naubasta Kalan in Allahabad District by University of Allahabad. Demonstrations were conducted on prawn hatchery technology and standardized methods for biocontrol of prawn disease for integrated giant freshwater prawn (Macrobrachium rosenbergii) culture. Trainings were conducted on scampi culture, breeding and seed production and microbial inputs were utilized for feed silage preparation using indigenous technology. Apiculture activity was introduced and disseminated to the target population and extension material was distributed to the beneficiaries.

Genetic Counseling

Tribal population of northeast region was screened for anaemia and nutritional deficiencies through a programme implemented at Vivekananda Institute of Medical Sciences, Kolkata. The tribes of Arunachal Pradesh, Assam and Tripura displayed high incidence of anaemia.

Awareness programme was organized for school children on inherited anaemia in pre-marital and pre-natal population. More than 1500 families from 80 villages were educated on anaemia, thalassaemia and hemoglobinopathies and counseling was provided to them. Leaflets were distributed for awareness of anaemia with special reference to Thalassaemia in local languages.

A project on molecular epideomiology of alcoholism was implemented jointly by Centre for Cellular and Molecular Biology, Hyderabad and Tribal Research Centre, Udagamandalam, Nilgiri District to study Kota tribe population in Nilgiri hills. General health awareness camps were arranged in seven villages for periodical medical check up. Blood samples were collected from 620 individuals and clinical data on health profile revealed that the disease morbidity is more for hypertension, heart related problems and diabetes. Genealogies indicated that the alcohol drinking habit is running in the families for the past two to three generations. Study undertaken on the health awareness camps and medical camps were found beneficial in counselling the tribal population.

Floriculture

Over 180 women from Shillong, Itanagar and Chandigarh were trained in cultivation of commercially important and hybrid species of orchids. A total of 118 women of Shillong have raised 6 species *viz.* Dendrobium fimbriatum var. oculatum, D. longicornu, D.lituiflorum, C. pendulum, C. aloifolium and C. giganteum in their backyard. These plants have started blooming. The beneficiaries have started selling their harvest in local market. Through S&T Council, Bhopal 200 women were trained in cultivation of gladiolus, marigold, mogra, rose and tuberose.

Food Processing

At College of Horticulture, Vellanikara, Thrissur, 88 women were trained in various aspects of entrepreneurship in preparation of cocoa products including molding, packaging and marketing. Twenty-two women have already set up their production unit and are selling their products in local market.

In another project at Fisheries College, Tuticorin, 500 women were trained in cook-chill technology for seafood and various parameters for quality assurance *viz.* estimation of moisture content, bacteria load, presence of bacterial species, the nutritive value, shelf-life etc. The trained women were grouped in 48 SHGs for establishing market for seafood products.

At the Central Food Technological Research Institute, Mysore data on biochemical and microbial analyses for nutritionally important factors like protein, amino acids, isoflavones, genistein, diadzein and methods for improvement of shelf life were collected on for 'Tofu' preparation. Also a byproduct of Tofu preparation, Okara, was found to be rich in protein and amino acids has been used as a food supplement. In another project at CFTRI, 17 women were trained in preparation of various food products of high nutritive value, methods of removal of anti-nutritional factors like phytic acid and trypsin inhibitor found in soya bean, better texture and improved shelf life. The food items included fermented products of soya bean and fruit pulp. Also 34 women were trained in spirulina production and its application as health food. Through CFTRI, technologies for rural bakery are being disseminated to rural women in three villages.

Poultry and Livestock Farming

The 'Action Research on Health and Socio Economic Development', Bolangir, Orissa, identified 100 women and trained them in poultry rearing. The beneficiaries were provided with parent birds of Carigold (dual purpose) and Vanaraja varieties.

The beneficiaries were able to earn Rs. 1000/per month through the sale of eggs and birds. The beneficiaries were also trained in feed formulation using local ingredients, management practices and entrepreneurship skill. The trained women formed SHGs and are approaching a bank for setting up their poultry farms. The market is established through state owned apex cooperative known as OPOLFED.

Through three projects supported at Tamil Nadu Veterinary & Agricultural Sciences University and its Extension Centres at Namakkal, Chennai and Tiruchirapally, over 1200 women were trained in poultry farming including feed formulation, farm management/ health care and preparation of various food items such as quail egg pickle and ready-to-eat food products etc.

- *Aquaculture*: Under a project at the North Eastern Hill University, Shillong, techniques were standardized for induced breeding for common carp and 4,00,000 fish seeds were reared up-to fries for distribution to the beneficiaries. Selected 92 women were trained in fish farming techniques including water quality testing, pond maintenance, induced breeding, harvesting and feed formulation using slaughter house waste. They were provided with fish fries for rearing in their own pond.

Over 470 women from Bhopal, Shimla, Solan and Bhubaneshwar were trained in polyculture fish farming. Interested women were also provided advance training in composite carp culture, integrated fish farming with duck/poultry, pond management and value added products including fish processing. These women are selling their products in local market and are earning Rs. 3000/per harvest/family.

Considering the opportunities in ornamental fish, a project was supported at the University of North Bengal. The identified women, 57 from Darjeeling and 54 from Jalpaigudi district were trained in rearing and breeding of guppy

(Poecilia reticulate) in cemented cisterns and angel (Pterophyllum scalaer) in ponds. The women are being trained in management skill and marketing.

Organic Farming/ Vermicompost

At the Sri Sathya Sai Institute of Higher Learning, Anantapur Campus, Andhra Pradesh, neem varieties with higher level of azadirachtin were identified in 5 villages. Thirty trees were selected and 40 women from two villages were trained in fruit collection, drying and storage. Specialised training on maintenance of neem processing gadgets, processing of neem fruits for production of cake, oil or powder, maintenance of aza-rich plantlets, hardening, harvesting and application of neem based biopesticide was provided to five resource persons

At the University of Agricultural Sciences, GKVK Campus, Bangalore, five villages of Bangalore and Kolar were surveyed for the availability of organic waste. The waste materials were mostly sugarcane trash, coconut coir pith and vegetable wastes. Demonstration programmes for 20 to 25 farmwomen from each identified village were organized. Six interested women were given onfarm training. The technique of Trichoderma enriched vermicomposting was perfected and demonstrated at University campus.

At MSSRF, Chennai, phosphate-solubilising local strains were collected and screened. Simultaneously several awareness programmes on application of biofertilizer were organized for the villagers of Rediarchatram block. About 300 interested farmers were trained in large-scale production of biofertilizer. Under the project support at Dangoria Charitable Trust Hyderabad, apart from training women in processing various items of food, 100 women from 13 villages were trained in production of vermicompost, neem-based biopesticide and their application in organic farming, About 100 vermicompost units were set up using the ring method. The beneficiaries produced over 130 quintals of compost. The beneficiaries were encouraged to raise nurseries of drumstick, papaya and creeper spinach at their backyard using vermicompost. The beneficiaries were able to sell excess vermicompost and planting material.

Horticulture

Through the Tamil Nadu Agricultural University, Coimbatore, 54 progressive farm women were identified for training on cultivation of hybrid rice. The women were provided with quality seed their cultivation over an area of approximately four acre.

The yield for the first harvest (December to May) was recorded at 2000 kg/ha. The beneficiaries were able to earn an additional income of Rs.20,000/

per acre/year through three crops in an year. At Supi farms, Uttaranchal non-traditional vegetable crops were grown on farmers' land in a small way. An herbal garden was established. Incidence of pests to various crops was evaluated and farmers were trained on production of vermicompost and application of neem based biopesticide in horticultural crops.

Mushroom Cultivation

In an attempt to empower rural women through transfer of mushroom cultivation technology 93, women were trained at Himachal Pradesh Krishi Pradesh Vishwavidyalaya, Palampur and 750 at Tropical Botanic Garden & Research Institute, Kerala in mushroom cultivation. In addition, 15 women were trained in spawn production at TBGRI. The beneficiaries were given 15 days training in cultivation of Oyster and Agaricus species using wheat straw. The first harvest of about 1000 kg was sold @ Rs. 16/kg. The beneficiaries were also trained in various recipies of mushroom. Market linkage has been established for their products.

Under the project support at Dangoria Charitable Trust Hyderabad, apart from training women in processing various items of food, 100 women from 13 villages were trained in production of vermicompost, neem-based biopesticide and their application in organic farming, About 100 vermicompost units were set up using the ring method. The beneficiaries produced over 130 quintals of compost. The beneficiaries were encouraged to raise nurseries of drumstick, papaya and creeper spinach at their backyard using vermicompost. The beneficiaries were able to sell excess vermicompost and planting material.

Sericulture

A total of 550 women were trained for various activities including the organic cultivation of mulberry & rearing of cocoon and value added products from waste at the Central Sericultural Research & Training Institute, Mysore and at Vidyasagar University, Midnapore, West Bengal. The beneficiaries are getting an additional income through the sale of value added products and surplus vermicompost.

Healthcare

Kidwai Memorial Institute of Oncology, Bangalore, Karnataka organized over 50 awareness camps on Carcinoma cervix to educate 2071 women for its prevention. Blood samples collected from 1830 women based on their medical history, 1812 were found negative and 18 positive cases were referred for treatment. A project on mental retardation including awareness generation among rual women, screening for early detection and genetic counseling was supported to Manovikas Bio-Medical Research. Over 7,000 families from three Gram Panchayats were surveyed. Ninety-one of these showing signs of mental retardation were shortlisted. The families were informed about the possible risk factors in future. Analysis of blood samples from these for genetic carrier showed only 4 cases of low level of fragile X chromosomes.

A project is supported at Saraswati Dental College, Lucknow for dental fluorosis. Water samples from 12 villages around Lucknow and Kanpur were analysed for fluoride content. Health camps were organized in 4 villages with a team of dental surgeon. A total of 2860 people were examined for dental fluorosis. The severely affected persons were advised for regular dental examination at SGPGI, Lucknow. They were also provided with calcium and cholicalciferol supplement. The blood samples from the severely affected boys and girls within the age group of 15 years and pregnant women were examined for their blood calcium level. Under another project supported at Mahavir Hospital And Research Centre, Hyderabad, Andhra Pradesh, blood lead levels in children and adults including men and women were measured to study environmental genotoxicity in women and children occupationally exposed to lead, a heavy metal.

CHALLENGE OF INDIAN AGRICULTURE

There are several concerns about Indian agriculture that we need to address. Even in the green revolution crops, rice and wheat, the Indian average yield ranks around 50th in the world. A weak agro-based industry, the problems of weak marketing, storage, transportation, credit support, etc, haunt us. Only 0.3% of agricultural GDP is spent on agricultural R &D. The population is growing at an alarming rate of 1.8% per annum. It will be around 1.3 billion by 2020 and would require 50% additional foodgrains. A total of 200 million people are still below the poverty line and have insufficient access to food. Declining resource base, degeneration of natural resources, including soil fertility, water environment, etc, are some other concerns.

Food production systems centre around a few states only. Out of 17 big states, only 7 are self-sufficient and surplus in foodgrains. There was a significant decline in per capita intake of calories and protein between 1975 and 1990.

Inspite of all these problems, there is plenty of good news. India harbours a vast diversity of basic natural resources. Tremendous bio-diversity of agriculturally important micro-organisms exists, which are not yet been harnessed. We have a variety of crops, including quality fruits, vegetables, foodgrains, and a number of plantation and commercial crops. Indian fisheries resource has spread over 8,100 km of coastline on the east and west, and provides access to marine resources and transportation.

Very large proportion of area of various food crops is under low productivity category. It can be increased with inputs of high class science. There is a scope to cultivate sizeable portion of waste lands through soil amendment and introduction of agroforestry. Several million hectares of saturated soil can be potentially exploited from rainfed lowlands of eastern India.

The India Exclusive Economic Zone (EEZ) of the marine sector, of about 2 million sq.km., has high potential harvestable yields. Only about 20% of the cultivated area is covered under tractor cultivation, great scope of making quantism quantum jump in production and productivity exists, changing consumption patterns offer scope for diversified agriculture. Trade liberalization provides opportunities to reach for outside markets, which were not accessible otherwise. Great gains are possible through agro-based industrialization and overall rural development.

The cultivators and farmers of India have a rich heritage of traditional agricultural wisdom on crop and livestock husbandry and fisheries; based on the principles of the conservation of natural resources and environment. India has an extensive network of extension channels for dissemination of technologies generated by research institutes.

India has a strong manufacturing capability and a large network of about 20,000 manufacturers and nearly one million village artisans for indigeneous production of all types of agricultural machinery. Therefore, inspite of several disadvantages, we have a great chance to leap forward, if we enforce the right policy initiatives and implement them.

Biotechnology and India Agriculture

We need to recongnise that, just as the food reequirements of today's population of nearly 6 billion people could not have been met by the technologies of the 1940's, we cannot assume that current practices will feed the population of 8 billion expected by 2020. New approaches are needed in addition to the continued improvement of existing methods of crop and animal husbandry and food processing.

The strategic integration of biotechnology tools into India agricultural systems can revolutionise Indian farming and usher in a new era. Genetic engineering is clearly the most revolutionary tool to impact agricultural research since the discovery of genetics by Mendel. Prior to genetic engineering, the exchange of DNA material was possible only between individual organisms of the same species. With the advent of genetic engineering in 1972, scientists have been able to identify specific genes associated with desirable traits in one organism. For example, a gene from bacteria, virus or animals may be transferred into plants to produce genetically modified plants having changed characteristics. This method therefore allows mixing of the genetic material from species that cannot otherwise breed naturally.

Genetic engineering can be vital for an agrarian country like India. It can help in minimising the crop damage through disease and pest resistant varieties, reducing the use of chemicals, enhancing stress tolerance in crop plants, thus permitting productive farming on unproductive lands, etc.

One could even extend the growing season of crops and minimise losses due to environmental factors on the one hand and increase the shelf life of fruits and vegetables, on the other, thus minimising losses due to food spoilage. We can expand the market vista and improve food quality. Biotechnology can also produce plants that possess healthy fats and oils, possess increased nutritive value, and create a whole range of higher value feeds. Biotechnology has even the power of producing biodegradable plastics, edible vaccines, etc.

It is only through the blending of the 'gene revolution' with our experience in the 'geen revolution' that we can reach our goal of 'evergreen revolution' and also 'nutritional revolution' The advantage of the gene revolution is that it is relatively scale neutral, benefiting big and small farmers alike, It is also environment friendly. Thus, it can be of great help to the smallest farmer with limited resources in increasing farm productivity through the availability of improved but powerful seed. It can also reduce a farmer's dependency on chemical inputs such as pesticides and fertilisers.

Modern biotechnology offers unlimited opportunity for enhancing genetic potential of crops and other commodities, management of biotic and abiotic stresses, bio-remediation and organic recycling. India is one of the main centres of agricultural biodiversity and its gene richness can greatly complement the developments in modern biotechnology. Likewise, new developments in GIS, remote sensing, and crop modelling, provide new opportunities for integrated management of natural resources. The revolution in informatics provides opportunities for sharing latest information for research and planning in a highly organised and efficient manner.

With the globalisation of economy, the opportunities for value addition and post harvest management are also immense. With clever blending of technologies, the India farmer can usher in new era of confidence and performance. Sir Francis Bacon had once said: *'It would be an unsound fancy to expect that things which have never yet been done, can be done except by methods which have never been tried'* In the new context, we have to review the new role of biotechnology.

Also, it is in this context that we have to view the rapid advances that modern biotechnology is making, including the area of transgenics. Let us focus on some of the developments in this area, since considerable interest and controversy has been created around the world today in this exciting area.

PRESENT STATUS OF AGRICULTURAL BIOTECHNOLOGY IN ASIA

Selected Country Experiences

Several governments in Asia are committed to the use of modern biotechnology in agriculture. They have devoted significant human and financial resources to this policy over the past two decades. The PRC accords high priority to biotechnology to increase food production and improve product quality in an environmentally sustainable manner. The PRC has moved quickly to adopt new biotechnologies, particularly genetically modified crops. The country is rapidly increasing its expenditure on biotechnology R&D.

Over 103 genes have been evaluated for improving traits in 47 plant species. New traits have been introduced and evaluated in field tests on rice, wheat, maize, cotton, tomato, pepper, potato, cucumber, papaya, and tobacco. A variety of traits have been targeted. They include resistance to diseases, pests, and herbicides, and quality improvement. Approximately 50 genetically modified varieties have been approved for environmental release, or small-scale field testing in the PRC.

A few new genetically improved varieties have been approved for largescale commercial production. The most widespread are new pest-resistant varieties of cotton that are being widely cultivated by farmers. These were grown commercially by approximately 3 million farmers on approximately 500,000 ha in 2000. Several new products are in the pipeline for potential commercialization. India has allocated large public resources towards human resources development and infrastructure in biotechnology.

In the early 1980s, the Government of India created a Department of Biotechnology to promote the use of new biotechnologies in industry, medicine, and agriculture. Current R&D efforts in India are directed towards increasing agricultural productivity, bioremediation in the environment, medical and industrial biotechnology, and bioinformatics. R&D priorities in agriculture include new regeneration techniques for the rapid multiplication of citrus, coffee, mangrove, vanilla, and cardamom.

Cardamom yield has increased 40 per cent through the use of tissue culture. There is substantial private sector participation in biotechnology in India, for example, in the seed sector, the veterinary products sector, and bioinformatics linked with the booming information technology sector. Further, organizations like the M.S. Swaminathan Research Foundation have developed innovative approaches such as the ADB-supported BioVillages programme that is fostering the growth of new income-raising technologies. Emphasis is on developing small-scale bioindustries for women. Indonesia has placed a high priority on biotechnology over the past 15 years. The Government has designated three National Biotechnology Centers to coordinate R&D in agriculture, medicine, and industrial microbiology.

Applications of biotechnology to agriculture are primarily the responsibility of the Agency for Agricultural Research and Development (AARD). A National Committee on Biotechnology advises the minister in developing guidelines for government policy in the promotion of biotechnology. In recent years there has been an extensive training programme within Indonesia and abroad to upgrade skills of scientists involved in biotechnological research. In the 1980s, a major World Bank loan of over $100 million financed the creation of three inter-university centers for agricultural, medical, and industrial biotechnology.

More recently, a current World Bank loan is financing facilities for agricultural biotechnology within AARD. Crop improvement efforts using modern biotechnology started in Pakistan in 1985, when a training course was held on recombinant DNA. Work is now concentrated on chickpea, rice, and cotton. Field evaluation is hampered by lack of biosafety regulations. There is some private investment in R&D of agricultural biotechnology. The government controls testing, multiplication, distribution, and biosafety issues for genetically modified crops.

Pakistan lacks firm policy and regulations regarding intellectual property rights (IPR) and patents involving biotechnology, and biosafety regulations for GMOs. The Philippines began its modern biotechnology programmes in 1980 with the creation of the National Institutes of Molecular Biology and Biotechnology in Los Banos, with a focus on agricultural biotechnology.

In 1997, the Agriculture Fisheries Modernization Act recognized biotechnology as a major strategy to increase agricultural productivity. The

Act provided a budget for agricultural biotechnology of almost $20 million annually for the next 7 years (4 per cent of the total R&D budget).

In 1998, the government funded these five high-level biotechnology research projects to develop:

1. New varieties of banana resistant to banana bunchy top virus and papaya resistant to ringspot virus,
2. Delayed ripening papaya and mango,
3. Insect-resistant maize,
4. Marker-assisted breeding in coconut,
5. Coconut oil with high lauric acid content.

Nongovernmentt organizations (NGOs) and other groups concerned about the safety of GMOs have been vocal in the Philippines. This is affecting field-testing and commercialization of transgenic crops. Products in the regulatory pipeline include new varieties of insect-resistant maize and insect- and disease-resistant rice. Thailand is focusing on the applications of biotechnology to traditional foods, fruits, and export commodities such as shrimp. R&D priorities are to increase production and reduce production cost on crops such as rice, cassava, sugarcane, rubber, durian, and orchids.

An early success in Thailand has been the development of new molecular diagnostics for the diagnosis and control of virus diseases in shrimp. These diseases cost the shrimp export industry over $500 million in lost production in 1996. The development and commercial use of the new diagnostics prevents the loss of an estimated 20–50 per cent of annual production, a saving of at least $100 million per year. There are also active agricultural biotechnology programmes in Bangladesh, Malaysia, Nepal, Sri Lanka, Singapore, and Viet Nam. B. Regional and International Programmes.

International Agricultural Research Centers International

R&D programmes using modern biotechnology are being conducted by the international agricultural research centers (IARCs), particularly the International Rice Research Institute (IRRI), International Maize and Wheat Improvement Center (CIMMYT), International Crops Research Institute for the Semi-Arid Tropics (ICRISAT), International Livestock Research Institute (ILRI), and International Service for National Agricultural Research (ISNAR).

The Center for International Forestry Research also uses biotechnology in the characterization of forest diversity in its Asian programme. The International Center for Living Aquatic Resources Management is using new technologies in the improvement of fisheries and aquaculture systems. ILRI is initiating a programme on Asian livestock improvement. The CGIAR centers invest approximately $30 million per year in modern biotechnology.

ISNAR Biotechnology Service

The ISNAR Biotechnology Service (IBS), with Japanese support, has been assisting selected Asian countries in developing human resources for managing biotechnology research programmes or institutions. IBS has developed specialized courses to enhance the capacity and competency of managers, focusing on strategy building, priority setting, managing biosafety and regulatory aspects, resource generation and deployment, product delivery, and information sharing as well as the establishment and management of linkages.

International Service for the Acquisition of Agri-biotech Applications

The International Service for the Acquisition of Agri-biotech Applications (ISAAA) is brokering public-private sector partnerships to facilitate technology transfer. It has current regional projects to increase the productivity of maize, papaya, and sweetpotato.

ISAAA is also establishing a new Asian knowledge center for crop biotechnology, based in the Philippines, in partnership with CAB International and the Southeast Asian Regional Center for Graduate Study and Research in Agriculture. 0Its purpose is to make available timely and balanced information on the risks and benefits of crop biotechnology to interested parties in Asia.

In doing so, it aims to provide training and study tours highlighting experiences not only in research but also with biosafety and intellectual property issues. The major external sources of assistance for agricultural biotechnology in the Asia/Pacific region are ADB, Australia, the Rockefeller Foundation, the United Nations, the United States Agency for International Development (USAID), and the World Bank.

Asian Development Bank ADB has made several strategic and innovative investments in agricultural biotechnology over the past decade.

These investments have been in the form of loans and technical assistance projects. a. Components of Agriculture and Science and Technology Loan Projects Several governments have requested ADB financial support for human resource development, laboratory facilities, and equipment for agricultural biotechnology programmes. These programmes have been integrating new applications of biotechnology into their conventional agricultural R&D

programmes. The applications include the use of new molecular diagnostics for pests and diseases and MAS for crop and livestock breeding. Such components are being supported under ongoing ADB projects in the Philippines, Sri Lanka, and Thailand. ADB has provided other loans and grants in closely related areas such as in integrated pest management in cotton under the Cotton Development Project in Pakistan. b. ADB Regional Technical Assistance Projects ADB has provided regional technical assistance grants for the development of the three regional crop biotechnology networks over the past decade.

The networks are the Asian Rice Biotechnology Network (ARBN), initiated in 1993, the Asian Maize Biotechnology Network (AMBIONET), initiated in 1998 and the recently established Asian Semi-Arid Crops Network.

These networks have been influential in developing capacity to use new techniques in crop breeding for the major cereal crops (rice, maize) and the crops important in the semi-arid regions (sorghum, pigeonpea, and groundnut).

The networks are managed by three IARCs (IRRI, CIMMYT, and ICRISAT) that work with national research institutes to implement the programmes on rice, maize, and semi-arid crops.

A recent evaluation of ADB's research investments through the international centers reported favourably on the achievements of the rice and maize biotechnology networks, and recommended further support for this type of research cooperation.

The networks provide a platform by which countries and the IARCs can collaborate in the use of the new tools of functional genomics to identify genes that control important traits such as drought and salinity tolerance. This will speed up the breeding of crop varieties with these characteristics that have been difficult to address through conventional breeding. There will be increasing opportunities for cooperation between the networks, as more knowledge is gained on the commonality of genes between species and their control.

Asian Rice Biotechnology Network

The rice growing countries in Asia recognized in the early 1990s that biotechnology could provide powerful new tools for rice improvement. These tools included tissue culture, wide hybridization, molecular markers, and genetic engineering. The ARBN was initiated in 1993 to provide a vehicle for collaborative research in these areas with universities and national rice research institutes.

The objectives of ARBN are to:

- Promote human resource and infrastructure development for biotechnology at selected national agricultural research systems (NARSs) institutes through joint research and training coordinated by IRRI, and Asian Development Bank ADB has made several strategic and innovative investments in agricultural biotechnology over the past decade. These investment
- Generate biotechnology tools and products for use by NARSs through IRRI research and infrastructure development. ARBN was established with funds provided by ADB and the German Government's Budenministerium fur Technische Zussamenarbeit, which has been supporting IRRI's development of biotechnological products of direct use to NARSs. ADB supports research and infrastructure development at NARSs with additional funding for training and shuttle research by NARSs scientists at IRRI.

BIOTECHNOLOGY IN THE GLOBAL COMMUNICATION ECOLOGY

Public debates about the safety of new products introduced in the market go back centuries and were often based less on science than on the politics of the time. Similarly, today, much of the debate about agricultural biotechnology is steered by myths and misinformation and not by science, writes Calestous Juma, professor and director of the Science, Technology and Globalization Project at the Kennedy School of Government, Harvard University.

The scientific community, with stronger support from governments, must do more to address science and technology issues with the public, he adds. Debates over biotechnology are part of a long history of social discourse over new products. Claims about the promise of new technology are at times greeted with skepticism, vilification or outright opposition—often dominated by slander, innuendo and misinformation.

Even some of the most ubiquitous products endured centuries of persecution. For example, in the 1500s Catholic bishops tried to have coffee banned from the Christian world for competing with wine and representing new cultural as well as religious values. Similarily, records show that in Mecca, in 1511 a viceroy and inspector of markets, Khair Beg, outlawed coffeehouses and the consumption of coffee.

He relied on Persian expatriate doctors and local jurists who argued that coffee had the same impact on human health as wine. But the real reasons lay in part in the role of coffeehouses in eroding his authority and offering alternative sources of information on social affairs in his realm. In public smear campaigns similar to those currently directed at biotech products, coffee was rumored to cause impotence and other ills and was either outlawed or its use restricted by leaders in Mecca, Cairo, Istanbul, England, Germany and Sweden.

In a spirited 1674 effort to defend the consumption of wine, French doctors argued that when one drinks coffee: "The body becomes a mere shadow of its former self; it goes into a decline, and dwindles away. The heart and guts are so weakened that the drinker suffers delusions, and the body receives such a shock that it is as though it were bewitched."

Butterfly Stories and Other Misinformation Tactics

Today similarly charged stories are told about genetically modified (GM) foods. In addition to claims about the negative impact of GM foods on the environment and human health, there are wild claims that associate GM foods with maladies such as brain cancer and impotence as well as behavioural changes. Some of these rumors are spread at the highest levels of government in developing countries.

The tactics employed in the debates are equally sophisticated. Critics of the technology have used instruments of mass communication to provide the public with information that is carefully designed to highlight the dangers they attribute to biotechnology. Advocates of biotechnology have often been forced to respond to charges against the technology and have only on rare occasions taken the initiative to reach out to the public.

This is particularly important because the general public does not readily understand the technical details of biotechnology products and so new communication approaches are needed. While advocates of biotechnology have often tried to rely on the need for scientific accuracy, critics employ rhetorical methods that are designed to invoke public fear and cast doubt on the motives of the industry.

The critics draw analogies between the "dangers" of biotechnology with the catastrophic consequences of nuclear power or chemical pollution. Indeed, they use terms like "genetic pollution" and "Frankenstein foods". Critics have also relied on the general distrust of large corporations among sections of the global community to make their case. In addition, they have made effective use of incidents, whose risks they have amplified. A muchquoted study by Cornell University researchers indicated that pollen from GM maize producing a Bt toxin killed the larvae of Monarch butterflies. This study was used to dramatize the impact of biotechnology on the environment. Subsequent published peer explanations of the limitations of the study and refutations of the conclusions did not change the original impression created by the critics of biotechnology.

In this case the real environmental issue was not if GM maize killed monarch butterfly larvae or not. The critical question was what impact the maize had on the environment compared to maize grown with chemical pesticides. It is the issue of relative risks that is important; not simply a single event examined outside the wider ecological context. But apparently, this kind of analysis would not serve the cause of critics.

It is notable that the critics of biotechnology have defined the rules of the debate in two fundamental ways. First, they have managed to create the impression that the onus of demonstrating safety lies with advocates of biotechnology and not on its critics. In other words, biotechnology products are considered unsafe until proven otherwise. Second, they have been effective in framing the debate in environmental, human health and ethical terms, thereby masking the underlying international trade considerations. By doing so, they have managed to rally a much wider constituency of activists who are genuinely concerned about environmental protection, consumer safety and ethical social values.

There is a general view that concerted efforts to promote public debate will improve communication and lead to the acceptance of biotechnology products. This may be the case in some situations.

But generally, the concerns are largely material and cannot be resolved through public debate alone. This is mainly because the root causes of the debate lie in the socio-economic implications of the technology and not mere rhetorical considerations.

It is possible that public debates will only help to clarify or amplify points of divergence and do little to address fundamental economic and trade issues. What then can be done under the circumstances, especially in relation to developing countries that are currently the target of much of the attention of advocates and critics of biotechnology? Operating in the new global communication ecology will require greater diversity of biotechnology products, an increase in the number of institutional players, enhanced policy research on life sciences and society, and stronger policy leadership.

Products Speak Louder than Words

Much of the debate on the role of biotechnology in developing countries is based on hypothetical claims with no real products in the hands of producers or consumers. Under such circumstances, communication and dialogue are not enough until there is a practical reference point. In other words, rebutting the claim of critics is not as important as presenting the benefits of real products in the market place.

This can best be achieved through collaborative efforts among local scientists, entrepreneurs, policy makers and legitimate civil society organizations. There is ample evidence to suggest that concerns over the safety of new products tend to decline as local participation and ownership in new technologies increase. Similarly, local participation in new technologies increases the level of trust in new technologies, thereby reducing the demand for non-science-based safety regulations.

For example, the word of a farmer from South Africa stating the positive impact of GM cotton on her welfare carries more weight than thousands of screaming press releases and empty headlines on both sides of the debate. This means that spreading the use of biotechnology not only promotes familiarity with the technology, but also generates the information needed to convince the public about the relevance and usefulness of the technology.

The broadening of the range of products is therefore a key aspect of the debate. This is particularly important in developing countries interested in using the technology to enhance local products and diversify their food base. Information on the development of drought-tolerant crops, for example, would be relevant to African countries while other regions might be interested in different products.

This view also suggests that general debates about the role of biotechnology are of little utility unless framed in the context of local needs and applications. The absence of a real stake in the technology creates a vacuum that is often filled with misinformation on the risks and benefits of the technology. Countries such as Kenya and South Africa that have their own biotechnology research programs have a more considered view of the technology.

Broadening the Constituency

Addressing the issue of biotechnology communication requires an improved understanding of the changing ecology of communication. The ecology includes a complex network of sources of information and opinion leaders as well as new communication tools that were hitherto not available to the public or advocacy groups.

In his days, Khair Beg was outraged to learn that coffeehouses had become an authoritative source of information on what was happening in his jurisdiction. Similarly, the Internet has become a more important communication tool than classical methods such as TV advertising. But unlike in the days of Khair Beg, the new communication ecology is global in character, making it possible to spread information widely and generate empathy among a diversity of activist organizations, including those that are unlikely to be affected by the technology.

These cyber-communities are built around a complex set of mailing lists that are not easily accessible. Correcting misinformation spread through such channels is difficult to do because of the complexity of the networks. While activists tend to use a diverse array of social movements to advance their cause, advocates have tended to focus on the use of centralized institutions whose impact is largely negligible in the modern communication ecology.

But creating the necessary diversity requires a broadening of the base of social movements that champion the role of science and technology in human welfare. One of the most important aspects of the biotechnology debate has been the role of the popular media. In Europe, for example, the media have played an important role in amplifying claims by critics or creating doubt about positions advanced by advocates of the technology. In contrast, support for the role of science does not usually have the polemical turn that newspaper editors relish.

The traditional view that science is based on immutable facts which can be passed on from an authority to the general public is being challenged by approaches that demand greater participation in decision-making. In other words, scientific information is being subjected to democratic practices.

The debate over biotechnology has pushed the frontiers of public discourse of technical matters. On the one hand, society is being forced to address issues that are inherently technical, and on the other, the scientific community is under pressure to accept non-technical matters as valid inputs to decision-making.

Thinking Ahead

Policy-oriented research institutions and think tanks play an important role in the war of words. It is notable that critics of biotechnology have made a considerable effort to create alliances with research institutions, including university-based departments. Much of the material used to question the safety of biotechnology often has the legitimacy of a research institution.

But non-partisan policy research on the role of biotechnology in society is largely lacking, and so those seeking to provide an alternative view have limited opportunities to obtain credible information. The lack of systematic research on the interactions between biology and society is a critical bottleneck in efforts to engage the public in dialogue on biotechnology.

This is particularly critical given the fact that advances in biology pose new ecological and ethical issues that are not associated with the physical and chemical sciences. For example, concerns over the inability to recall products once released on markets are more pronounced when dealing with the release of biological inventions into the environment.

Leading the Way

Much of the public debate is intended to influence government policy on biotechnology. In this regard, the capacity of governments to assess the available information and use it for decision-making is an essential element of the debate. Political leadership on biotechnology and the existence of requisite institutions of science and technology advice are an essential aspect of the governance of new technologies. Debates over new technologies will be more pronounced in the future, and governments will increasingly come under pressure to address these issues.

But science and technology advice will not be sufficient unless governments view science and technology as integral to the development process. In this regard, enhancing the capacity of leadership to address science and technology issues will contribute to the effective management of public debates over new technologies in general and biotechnology in particular. On the whole, the nature of emerging technologies—particularly those based on the life sciences—and the changing ecology of communication are making it necessary to rethink strategies for advancing the role of biotechnology in society.

The scientific community will need not only to demonstrate a clear sense of leadership, but also to adapt its communication methods to suit the growing complexity and diverse needs of the global community. In the final analysis, it is the range of useful products available to humanity from biotechnology that will settle the debate, not the hollow pronouncements of advocates and critics.

THE PATTERN IN DEVELOPED COUNTRIES

In the industrialized countries, a new pattern of biotechnological research funding has emerged. With the availability of property right protection of biotechnologies and prospects of vast markets for biotechnology products and the techniques, the bulk of the research is funded, carried out, and controlled by the private sector. For instance, since 1976 when GENENTECH, the first biotechnology company, was established, the new techniques have spawned a variety of industries that now comprise more than 400 start-up firms, more than 200 established firms that have diversified into biotechnology, and more than 200 supply firms in the United States alone.

The new biotechnology industry in the United States produced pharmaceutical, diagnostic tests and agricultural products worth close to US$2 billion in 1990. A similar trend is seen in Europe and Japan. It is estimated that about 60 percent of the biotechnology research and development funding in industrialized countries is from the private sector. Thus, the private sector has been a major force in enhancing the capability of these countries in this field.

Research institutions in the public sector are now generally required to raise a substantial part of their budget from non-government sources, via contractual research, licensing agreements and royalties. This is tending to increase secrecy over research findings and to hinder free scientific communication. University professors, researchers and government institution scientists are increasingly becoming entrepreneurial and are entering private industry.

Another important trend is that large multinational corporations are purchasing smaller seed and biotechnology companies and diversifying their holdings. This allows them to develop a package sale of chemicals, seeds and equipment. Heavy involvement of the private sector and market considerations greatly influence the topics and commodities chosen for research. Major crops, commodities and farming systems of great socio-economic importance to the developing world, but of little international market importance, do not figure in the biotechnology research agenda of industrialized countries. Furthermore, these countries are keen to reduce their production costs, increase the productivity, quality and value of their products and, thus, improve their overall competitiveness in the world market.

The Patterns in Developing Countries

Biotechnology facilities are being established in most developing countries. However, the level of research, development and use of biotechnology for agriculture, forestry and fisheries in the developing countries is generally far below the level in the industrialized countries. Among developing countries, the status varies considerably.

A few, such as Brazil, Mexico, India, China and The Republic of Korea, have sought to gain full scientific and technological capacity, especially in agricultural biotechnology. Others, such as Indonesia, Malaysia, the Philippines and Thailand and a few countries of Latin America, have built the capacity to apply biotechniques and develop biotechnologies useful for their agriculture and food industries. The participation of the private sector in gaining biotechnological capacity is not significant in most of these countries.

Many developing countries have inadequate funding, poor human resources and limited access to information, resulting in a relatively low-level capacity for research and technology development and exploitation, especially in the field of modern biotechnology research, which is costly and requires highly trained personnel.

Most developing countries are vague as to their immediate aims in agricultural biotechnology. Few have appropriate proprietary-right protection systems or mechanisms to increase their access to protected techniques and products. Furthermore, there is negligible involvement of the private sector, which accentuates the problem of insufficient attention to biotechnology.

One of the major constraints on biotechnological development in developing countries is the poor quality and extent of higher education in frontier sciences, especially molecular biology. Besides, there are no or very weak links between universities and research institutions to reinforce each other and to utilize synergistically the scarce trained personnel resources.

The university experts play a negligible role in national policy formulation, including that on biotechnology. Furthermore the universities are generally not oriented for efficiency in commercialization and are therefore not able to market biotechnology products of their own or others' inventions.

Some international and donor-country public sector institutions strive to ensure that attention is given to solving, through biotechnology, the food and agriculture problems of developing countries. Among this group are the International Agricultural Research Centres (IARCs) located or operating in developing countries.

Capacity exists in several of these centres for utilization of advances in biotechnologies to solve certain problems of plant production and protection and of livestock production and health, and capacity is being built in others. For example, the International Rice Research Institute (IRRI) is incorporating modern molecular biology in its rice research activities. Its programme is linked closely to the Rockefeller Foundation Biotechnology Network. The World Bank, the International Service for National Agricultural Research

(ISNAR) and the Australian Government undertook a major study on the likely impact of modern biotechnology on agriculture. This study is being followed up by the participants by an expansion in the World Bank lending for biotechnology and by ISNAR's establishing an Intermediary Biotechnology Service to provide advisory services to national agricultural research systems.

Applications, Impact and Potential

Biotechnological research and development are moving at a very fast rate. For instance, three years ago, transgenics in rice, especially in japonica rices, were considered rather difficult because of the problem of regeneration through protoplast culture. But today, a large number of transgenics both in indica and japonica types are available and are being tested at various stages. It is therefore difficult to predict the potential and impact of a given technology beyond five years or so. The current level of technology and refers to the developments likely to occur in the next three to five years.

The potential of genetic engineering in medicine has received much attention. The potential for advances in agriculture, forestry and fishery are similarly bright. Biotechniques are already being used to create new strains of crop plants, new plant and animal diagnostic products, animal vaccines, biological pesticides and herbicides, other biological control agents, and modifications in domestic animals used for food production. In some cases, applications are being held in check by the need for still more research to ensure that there are no harmful effects (or, if there are, they are greatly outweighed by the benefits) and by the slow pace of evolution of regulations governing the release of genetically modified organisms or products for general use or acquisition.

Biotechnological research, especially genetic engineering, on problems of field and tree crops, forest species and fisheries has a relatively short history. The knowledge of biological aspects of species and ecosystems needs to be expanded in may cases before new biotechnologies can be applied effectively.

Even where transfer of one or more genes can be achieved reliably, expression of the transferred gene(s) may not occur in the expected way. The search for and identification and cloning of useful genes, effective vector systems, methods of gene transfer and promoter mechanisms should be intensified.

A perspective on both the difficulties and intensity of the research effort suggests that diagnostic tools will speed solutions in the next five to ten years to certain disease problems in several crop, forest, animal and fish species, such as black sigatoka of banana and plantain, and virus/viroid diseases of coconut and rice. Improved tolerance to certain physical stresses, such as heat tolerance in potatoes, may also be achieved, but prospects for improved resistance of varieties and clones to most abiotic stresses are longer term.

Crops

Modern biotechnologies can add greater precision and speed to plant breeding. Transgenics have already been reported in 40 crop plants, including maize, rice, soybean, cotton, rapeseed, potato, sugar beet, tomato, potato and alfalfa, but the new varieties are yet to be used commercially. Near-future opportunities for commercial exploitation include vegetables and fruits (potato, tomato, cucumber, cantaloupe and squash), followed by legumes (alfalfa) and oilseed crops (oilseed rape). A good number of the transgenics are herbicide-resistant plants whose widespread use is somewhat controversial.

Wide use is currently made of tissue culture techniques for micropropagation of elite clones and for freeing planting materials of pathogens.

Monoclonal antibodies are also in use as diagnostic aids in the detection and identification of viruses and viroids. Another culture and microspore culture giving rise to haploids are being used in variety improvement to facilitate and accelerate breeding. Molecular maps and markers are being widely used to identify genes of interest to accelerate conventional breeding programmes. Efficient biological nitrogen fixation systems and strains for efficient utilization of soil nutrients are being genetically engineered. Other long-term objectives are the genetic manipulations of photosynthesis pattern and the production of hybrid seed through apomixis. A very distant possibility is providing nitrogen fixation capacity to cereals.

Livestock

Among agricultural and allied fields, animal production and health have benefited the most from biotechnology, although practical use of transgenic livestock is only a future possibility. Wide use of monoclonal antibodies for efficient diagnostics, leading to safe and specific treatments of animal diseases, is a major breakthrough.

Through genetic engineering, vaccines for the prevention of viral, bacterial, and parasitic animal diseases are rendered more effective and safe. Tailored vaccines exist for pig scours, chicken bursar disease and cattle tickborne diseases. Pathogen-specific vaccines are attractive goals. Other interesting possibilities are endocrine-directed vaccines to stimulate twinning in beef cattle, immunocastration, livestock growth rate increase and vaccines that compensate for various stress-induced production losses.

Advances in genetic engineering will also facilitate the production of male-only populations of screwworms, tsetse flies, ticks and various other ectoparasites for use in the sterile release technique of control or eradication. Furthermore mammalian tissue culture may replace whole animals in the 1990s for toxicity testing of certain chemicals. The culture technique can also be exploited for studying and analyzing pesticide metabolism and for herbicide pre-screening. In vitro fertilization and embryo sexing techniques have considerably increased the use of embryo transfer techniques for cattle breeding and trade.

The value of the approach will be further enhanced if embryo cloning techniques can be reliably employed. Microbial and enzymatic treatment of roughages and genetic engineering of rumen bacteria both have great potential to improve animal nutrition. Growth hormones can be produced by genetically engineered microorganisms in quantities and at the low cost necessary for widespread use to speed and increase milk and lean meat production. Biotechnological tools (embryo culture, gene cloning, etc.) may also be used for conservation of genetic resources.

Forestry

Successful regeneration through micropropagation or somatic embryogenesis has been achieved for about 100 forest species although, for most, considerable development work would be required before commercial propagation could be contemplated. In breeding programmes incorporating clonal testing, inclusion of a micromultiplication phase may facilitate more rapid deployment of superior genotypes than that afforded by sequential multiplication by cuttings.

This will particularly be the case when gene mappling techniques are sufficiently advanced to permit accurate identification of superior genotypes without replicated clonal testing in the field, especially within a breeding line where substantial genetic disequilibrium exists. Furthermore, gene mapping may allow eventual identification of valuable genes, including those that contribute to quantitative traits.

As regards the application of genetic engineering, many major traits of commercial importance to forestry are under polygenic control, and much remains to be learned about the operation of the genes involved before this technique can have a major impact for these characters. An earlier application of genetic engineering in some forestry operations may be the introduction of genes, some already available, known to confer insect of disease resistance. Other applications of biotechnology with obvious value for forestry, but as yet not widely supported by experimental successes, include: in vitro manipulation of the maturation state, *e.g.* the promotion of early flowering to reduce generation intervals, in vitro selection for traits such as disease resistance and tolerance to salinity; and the use of haploid cultures.

Fisheries

Two broad areas of biotechnology exist within the fisheries sector, namely natural products biotechnology (including mostly marine species); and aquaculture biotechnology. Commercially valuable products such as pharmaceutical compounds, pigments, oils, sterols, alginates and agarose are being extracted from micro-and macro-algae in many parts of the world.

Aquatic invertebrates are currently being screened for biologically active compounds that may have, inter alia anti-carcinogenic or antiviral properties, whereas marine bacteria are currently utilized in treatments of oil spills and holding tanks in tanker ships.

Within aquaculture, induced increases in the chromosome complement (polyploidy) of commercially important species such as salmon and oysters have increased the growth and marketability of these species. Genetically engineered microbes can be used to produce fish growth hormones, which might then be used to improve feed conversion and growth rate. Synthetic reproductive hormones are produced commercially and used to regulate fecundity, breeding cycles, growth rates and sex determination in certain cultured species.

In attempts to increase desirable culture qualities such as growth rate, disease resistance, temperature tolerance and marketability, transgenic fishes containing introduced genes from other species have been produced on an experimental scale. The application of advanced biotechnology to the fisheries sector is a relatively recent practice. Therefore, continued research can be expected to reveal additional commercially viable applications and products within this sector.

BIOTECHNOLOGY INNOVATION SYSTEM IN INDIA

The government plays an important role in building the National Innovation System (NIS) by setting up institutions (public research institutions such as universities, labs etc), developing policies that support innovation and providing regulatory frameworks that can have a considerable influence on sectoral development. In the context of the developing countries, the role of the government becomes more prominent as the majority of the resources for the scientific research and development are being channelised or governed by the government. In case of India, nearly 80% of the gross expenditure for research and development is being incurred by the Government while in case of the industrialized nations such as Japan, USA and UK the Industry spends nearly 75%, 64% and 44% respectively.

Though there has been a significant increase in the country's Gross National R&D expenditure in absolute terms from Rs.3974 Crore in 1990-91 to Rs.16199 Crore in 2000-01 to Rs.18000 Crore in 2002-03, the share of the government in national R&D expenditure has remained unchanged. However, it is interesting to note that internationally all the governments perform the in-house R&D function, including even the highly private-sector-oriented governments such as the US and other OECD countries.

A study of the Biotechnology innovation system of the four global leading countries namely USA, UK and Japan and Germany further supports the active involvement of Government R&D in furthering the growth of the Biotechnology sectoral innovation. Biotechnology is being characterised by multidisciplinary fields, high dependence on science and are emerging from industries having large R&D base, high degree of qualified manpower and having close links or interaction with public research institutes as well as universities and thus faces a broad set of different policy interventions by the government underlying the National Innovation System (NIS).

The National System of Innovation (NSI) perspective is a convenient framework to understand the process of innovation occurring in an economy and especially within the manufacturing sector. It focuses on the country-specific factors influencing the process of innovation and economic growth. The NSI approach has evolved over the years both analytically and empirically with the contributions from Freeman, Lundvall and Nelson, Edquist (1997, 2002). NSI has became popular as an analytical framework to study innovation at the national, regional and even Sectoral level.

Though there are several definitions of NSI but that best fit in to the present context is that of Metcalfe (1995). NSI is defined by Metcalfe (1995) as a "set of distinct institutions which jointly and individually contribute to the development and diffusion of new technologies and which provides the framework within which governments form and implement policies to influence the innovation process. As such it is a system of interconnected institutions to create, store and transfer the knowledge, skills and artefacts which define new technologies".

NSI framework is comprised of various components such as the government, independent research institutes, firms and higher education system etc and their interrelationships. Innovation or government policy instruments and institutions play a very important role in cementing the relationship among the various components for the success of NSI. An application of this framework shall help in analysing the systemic failure that hamper the generation of innovations. Public policy can be applied to correct such failure. There are several NSI studies on Biotechnology in the context of the developed world but the studies in the developing countries are rather limited.

Biotechnology NSI in India has evolved over time and has been instrumental in creating infrastructure and capabilities in promoting biotechnology innovation in the country. There are 241 institutions and 321 firms engaged in various dimension of biotechnology in the country.

Indian Biotechnology industry is now among the top 12 countries in the world and like its global counterpart, is dominated by the healthcare sector. The recently released "Draft National Biotechnology Strategy 2005" by the Indian Government is a step forward in the direction of promotion and development of biotechnology to withstand global challenges and yet, solve the local healthcare needs of the country. Studies on biotechnology in India by and a recent empirical analysis of the health innovation system of seven developing countries further supports the role of the Governments in providing the sustainable environment for the successful development of health biotechnologies.

According to Acharya *et. al.* government in developing countries is one of the key actors in prioritizing, implementing new technologies and promoting innovation to address their specific health and development needs. The governments of the countries where biotechnology innovation system has been successful in perform or support the following functions/activities such as knowledge support, commercialization of research, collaborative research, financial support, regulation, IPR and resolving ethical issues etc.

Thus it would be interesting to analyse the role of the government in the promotion of the biotechnology in the country with special focus on the healthcare sector.

Hence, the present study traces the role of government or national approaches in terms of policies, institutions and instruments in the promotion of biotechnology innovation system in India with a special focus on the healthcare biotechnology sector. A systemic framework similar to the EPHOITE (2003) project is being adapted to examine the direction of public policies, institutions and instruments in addressing the issues of knowledge support, commercialization of research, collaborative research, financial support, regulation, IPR and ethical issues in the area of healthcare biotechnology in Indian context.

The significant features of national biotechnology innovation system with focus on healthcare biotechnology with the help of various S&T indicators. The various government initiatives in terms of various policies, institutions and instruments in the framework of NSI while the lessons learnt along with the policy implications and challenges for the growth of healthcare biotechnology innovation.

MODERN AGRICULTURAL BIOTECHNOLOGY

In 1973, Cohen and Boyer transferred a gene from one organism into another. In 1982, the first biotech plant, an antibiotic resistant tobacco, was developed. In January 1983, at a meeting of genetic researchers in Miami, three different teams reported success in using Agrobacterium tumefaciens, a bacterium, to carry new genes into plant cells, heralding the dawn of modern agricultural biotechnology. Agrobacterium tumefaciens, described as a "natural genetic engineer", splices its own genes into host plant cells. This pathogenic bacterium was now converted into a pack mule, to carry new, foreign genes into plant cells, minus the disease and this became the most common means of producing Genetically Engineered Organisms (GEOs).

Field tests for GE crops resistant to pests and pathogens were first conducted in the US, in 1985. A co-coordinated framework for the regulation of GEOs was established and the first GE tobacco was released, in 1986 . The US Department of Agriculture published guidelines for field trials of GE crops in 1991.

On approval from the Food and Drug Administration, Flavor Saver, the first GE tomato, with a longer shelf life, was on the US markets in 1994. During 1995-96, GE soybean, corn and cotton were approved for commercialization, in the US.

A number of GE crops, developed for pest, pathogen and herbicide resistance, are now commercially cultivated in several countries. Rice with pro-vitamin A, higher iron content, or human milk proteins and potatoes with high protein content are in various stages of development. Tobacco plants producing functional human haemoglobin and bacteria that produce human insulin have been developed, as well as plants with vaccines against rabies and other viral diseases. Food grain crops that withstand drought and salinity are high priority research, and so are those for high yield. A GE tobacco plant detoxifies soils contaminated by explosive residues, providing solution to a frustrating environmental problem in countries ravaged by armed conflict.

Now more than 70 biotech agricultural crops that have been approved for use in North America, including varieties of soybeans, cotton, canola, corn, potatoes, squash, tomatoes and papaya. About six million farmers in some 17 countries now cultivate GE crop on about 125 million acres, a 30 fold increase over 1996. By end of the year 2002, six GE crops planted in the US (soybeans, corn, cotton, papaya, squash and canola) produced an additional four billion pounds of food and fibre on the same acreage, improved farm income by US $ 1.5 billion and reduced pesticide use by 46 million pounds. In 2003 in the US, 80 per cent of soybean acres will be planted with biotech varieties.

A number of activist groups vehemently attack GE technology and its products on grounds of safety to humans and the environment, and costs of technology transfer and its reach to the needy. Products of agricultural biotechnology bear the brunt of this ill-informed, unscientific and prejudiced onslaught much more than GE products related to health or industry.

In 2001, the European Community released results of a 15-year study, costing US $ 64 million, and involving more than 400 research teams and 81 projects. This report concluded that GE products pose no more risk to human health or the environment than conventional crops. So far, extensive and intensive research on the probable risks of GE technology has not brought out any adverse effects and none of the fears expressed by anti-tech activists were proved even marginally. Nevertheless, caution and examining issues case by case, is the watch word of technologists, who are aware of their responsibilities.

For a number of products, technology transfer is free of costs for developing countries, as for example Golden Rice, the rice with pro-vitamin A. It is the responsibility of the Governments of the respective countries to bear the costs of developing local varieties and to reach the products to the needy at an affordable cost. In India, three GE varieties of pest resistant cotton were approved for commercialization, a year ago. Approval for other GE varieties of cotton and GE mustard was deferred twice by the Genetic Engineering Approval Committee (GEAC), the highest authority on the issue.

In India, the level of public awareness of the realizable benefits and probable risks of GE products is abysmally low. The functioning of the GEAC leaves much to be desired. Taking advantage of this hazy situation, mixing up economic, social and political issues with science, and even using such vagaries of nature as the severe drought, several groups of vested interest have created mistrust, confusion and scare. Trashing a very promising technology this way does not augur well for the future of Indian agriculture. This results only in denying the benefits of technology to the farmers and consumers.

The next 20 years of plant biotechnology is expected to improve quality of life, through improved foods, pharmaceuticals, new industrial materials and a better environment. These benefits should reach the people of the developing world, who need them most. For India to benefit from these developments, it is urgent and essential that the public is educated on the realizable benefits and probable risks related to GE products. We also need to reorganize the mechanism of regulating GE products by providing for an expedient, transparent and responsible authority. The media have a very important role to play in this regard. Only an informed and reassured public can make viable choices. Every effort should be made to provide such an opportunity to the consumer.

3

Urban Agricultural Planning and Organic Farming Management

IMPLEMENTATION OF URBAN AGRICULTURE

Creating a community-based infrastructure for urban agriculture means establishing local systems to grow and process food and transfer it from farmer (producer) to consumer. To facilitate food production, cities have established community-based farming projects. Some projects have collectively-tended community farms on common land, much like that of eighteenth-century Boston Common. One such community farm is the Collingwood Children's Farm in Melbourne, Australia. Other community garden projects use the allotment garden model, in which gardeners care for individual plots in a larger gardening area, often sharing a tool shed and other amenities. Seattle's P-Patch gardens use this model, as did the South Central Farm in Los Angeles. Independent urban gardeners also grow food in individual yards and on roofs.

Roof gardens allow for urban dwellers to maintain green spaces in the city without having to set aside a tract of undeveloped land. There are a growing number of projects worldwide that seek to enable cities to become 'continuous productive landscapes' through the networked cultivation of vacant urban land and temporary or permanent 'kitchen gardens'. Food processing on a community level has been accommodated by centralizing resources in community tool sheds and processing facilities for farmers to share. The Garden Resource Program Collaborative based in Detroit has cluster tool banks. Different areas of the city have toolbanks where resources like tools, compost, mulch, tomato stakes, seeds, and education can be shared and distributed with the gardeners in that cluster. Detroit's Garden Resource Program Collaborative also strengthens their gardening community by providing to their members transplants; education on gardening, policy, and food issues; and by building connectivity between gardeners through workgroups, potlucks, tours, field trips, and cluster workdays.

Farmers' markets, such as the farmers' market in Los Angeles, provide a common land where farmers can sell their product to consumers. Large cities tend to open their farmers markets on the weekends and one day in the middle of the week. For example, the farmers' market of Rue Richard Lenoir in Paris, France, is open on Sundays and Thursdays. However, to create a consumer dependency on urban agriculture and to introduce local food production as a sustainable career for farmers, markets would have to be open regularly. For example, the Los Angeles Farmers' Market is open seven days a week and has linked several local grocers together to provide different food products. The market's central location in downtown Los Angeles provides the perfect interaction for a diverse group of sellers to access their consumers.

Havana, Cuba

Due to the shortage of fuel and therefore severe deficiencies in the transportation sector a growing percentage of the agricultural production takes place in the so-called urban agriculture. In 2002, 35,000 acres (140 km^2) of urban gardens produced 3.4 million tons of food. In Havana, 90% of the city's fresh produce come from local urban farms and gardens. In 2003, more than 200,000 Cubans worked in the expanding urban agriculture sector.

Mumbai, India

Economic development in Mumbai brought a growth in population caused mainly by the migration of laborers from other regions of the country. The number of residents in the city increased more than twelve times in the last century. Greater Mumbai, formed by City Island and Salsette Island, is the largest city in India with a population of 16.4 million, according to data collected by the census of 2001. Mumbai is one of the densest cities in the world, 48,215 persons per km^2 and 16,082 per km^2 in suburban areas. In a scenario like this, urban agriculture seems unlikely to put in practice since it must compete with real estate developers for the access and use of vacant lots. Alternative farming methods have emerged as a response to scarcity of land, water, and economic resources employed in UPA.

Dr. Doshi's innovative techniques on farming: Dr. Doshi's city garden methods are revolutionary for being appropriate to apply in reduced spaces as terraces and balconies, even on civil construction walls, and for not requiring big investments in capital or long hours of work. His farming practice is purely organic and is mainly directed to domestic consumption. His gardening tools are composed of materials available in the local environment: sugarcane waste, polyethylene bags, tires, containers and cylinders, and soil.

The containers and bags (open at both ends) are filled with the sugarcane stalks, compost, and garden soil, which make possible the use of minimal quantity of water if compared to open fields. Dr. Doshi states that solar energy can replace soil in cities.

He also recommends the idea of chainplanting, or growing plants in intervals and in small quantities rather than at once and in large amounts. He has grown different types of fruit such as mangos, figs, guavas, bananas, and sugarcane stalks in his terrace of 1,200 sq ft (110 m^2) in Bandra. The concept of city farming developed by Dr. Doshi consumes the entire household's organic waste. He subsequently makes the household self-sufficient in the provision of food: 5 kg of fruits and vegetables are produced daily for 300 days a year.

City farm at Rosary High School, Dockyard Road Mumbai: The main objectives of this pilot project were to promote economic support for street children, beautify the city landscape, supply locally produced organic food to urban dwellers (mainly those residing in slums), and to manage organic waste in a sustainable city.

The project was conducted in the Rosary School, in Mumbai, with the participation of street children during 2004. A city farm was created in a terrace area of 400 sq ft (37 m^2). The participants were trained in urban farming techniques. The farm produced vegetables, fruits, and flowers. The idea was spread the concept of city farm to other schools in the city.

Mumbai Port Trust (MBPT): The central kitchen distributes food to approximately 3,000 employees daily, generating important amounts of organic disposal. A terrace garden created by the staff recycles ninety percent of this waste in the production of vegetables and fruits. Preeti Patil, who is the catering officer at the MBPT explains the purpose of the enterprise:

"Mumbai Port Trust has developed an organic farm on the terrace of its central kitchen, which is an area of approximately 3,000 sq ft (280 m^2). The activity of city farming was started initially to dispose of kitchen organic waste in an ecofriendly way. Staff members, after their daily work in the kitchen, tend the garden, which has about 450 plants."

Beijing, China

Beijing's increase in land area from 4,822 km^2 in 1956 to 16,808 km^2 in 1958 led to the increased adoption of peri-urban agriculture. Such "suburban agriculture" led to more than 70% of non-staple food in Beijing, mainly consisting of vegetables and milk, to be produced by the city itself in the 1960s and 1970s.

Recently, with relative food security in China, periurban agriculture has led to improvements in the quality of the food available, as opposed to quantity. One of the more recent experiments in urban agriculture is the Modern Agricultural Science Demonstration Park in Xiaotangshan.

Bangkok, Thailand

In early 2000, urban gardens were started under the direction of the NGO, Thailand Environment Institute (TEI), to help achieve the Bangkok Metropolitan Administrations (BMA) priority to 'green' Thailand. With a population of 12 million and 39% of the land in the city vacant due to rapid expansion of the 1960's-80's Bangkok is a test bed for urban gardens centered on community involvement. The two urban gardens initiated by TEI are in Bangkok Noi and Bangkapi and the main tasks were stated as:

- Teach members of the communities the benefits of urban green space.
- Create the social framework to plan, implement, and maintain the urban green space.
- Create a process of method to balance the needs of the community with the needs of the larger environmental concerns.

While the goals of the NGO are important in a global context, the community goals are being met through the work of forming the urban gardens themselves.

In this sense the creation, implementation, and maintenance of urban gardens is highly determined by the desires of the communities involved. However, the criteria by which TEI measured their success illustrates the scope of benefits to a community which practices urban agriculture. TEI's success indicators were:

- Establishing an Urban Green Plan
- Community Capacity Building
- Poverty Reduction
- Links with Government
- Status of Women<
- Developing a Model for Other Communities

Evan D.J. Fraser wrote in the article Urban Ecology in Bangkok Thailand that although the project was initiated to serve the environmental needs of the city it quickly illustrated the positive side effects of urban agriculture. "In many ways, the urban environment became a lens through which communities re-evaluated their own relationship with the city, the impact of urbanization in a global context, and how small groups can exert some control over the shape of their neighbourhoods."

Shenzhen, China

Traditionally Chinese cities have been known to mix agricultural activities within the urban setting. Shenzhen, what was once a small farming community is now a fast growing metropolis due to the Chinese government designation as an open economic zone. Due to large and growing population in China, the government supports urban self-sufficiency in food production. Shenzhen's village structure, sustainable methods, and new agricultural advancements initiated by the government have been strategically configured to supply food for this growing city.

Village Structure, Methods, and New Advancements

- The city farms are located about 10 km from city center in a two-tier system. The first tier approached from city center produces perishable items. Located just outside these farms, hardier vegetables are grown such as potatoes, carrots, and onions. This system allows produce to be sold in city markets just a few short hours after picking.
- Another impressive method used within Chinese agriculture and aquaculture practice is the mulberry-dyke fish-pond system, which is a response to waste recycling and soil fertility. This system can be described as, "mulberry trees are grown to feed silkworms and the silkworm waste is fed to the fish in ponds. The fish also feed on waste from other animals, such as pigs, poultry, and buffalo. The animals in turn are given crops that have been fertilized by mud from the ponds. This is a sophisticated system as a continuous cycle of water, waste and food…with man built into the picture"
- As population grows and industry advances the city tries to incorporate potential agricultural growth by experimenting in new agricultural methods. The Fong Lau Chee Experimental Farm in Dongguan, Guangdong has worked with new agricultural advancements in lychee production. This farm was established with aspirations of producing large quantities and high quality lychees, by constantly monitoring sugar content, and their seeds. This research, conducted by local agricultural universities allows for new methods to be used with hopes of reaching the needs of city consumers.

Government Involvement

- However due to increased levels of economic growth and pollution some urban farms have become threatened. The government has been trying to step in and create new technological advancements within the agricultural field to sustain levels of urban agriculture.
- "The city plans to invest 8.82 billion yuan in 39 agricultural projects, including a safe agricultural base, an agricultural high-tech park, agricultural processing and distribution, forestry, eco-agricultural tourism, which will form an urban agriculture with typical Shenzhen characteristics" in conjunction with this program the city is expected to expand the Buji Farm Produce Wholesale Market.
- According to the Municipal Bureau of Agriculture, Forestry and Fishery the city will invest 600 million yuan on farms located around the city, with hopes of the farms to provide "60 percent of the meat, vegetables and aquatic products in the Shenzhen market".
- There has also been an emerging trend of going green and organic as a response to pollution and pesticides used in farming practices. Vegetable suppliers are required to pass certain inspections held by the city's Agriculture Bureau before they can be sold as "green".

New York, NY

In New York City many low-income residents suffer from high rates of obesity and diabetes, limited sources of fresh produce and available, undeveloped land. The City and local nonprofit groups have been providing land, training and financial encouragement, but the impetus has in urban farming has really come from the farmers, who often volunteer when their regular work day is done.

Some urban gardeners have used empty lots to start community or urban garden. However, the soil must be tested for heavy contamination in city soil because of vehicle exhaust and remnants of old construction. However, studies have found that such ground can be cultivated as long as the pH is kept neutral. The City also has a composting program, which is available to gardeners and farmers. One group, GreenThumb, provides free seedlings. Another program, the City Farms project operated by the nonprofit Just Food, profits courses on growing and selling food.

Benefits of Urban Agriculture

The benefits that UPA brings along to cities that implement this practice are numerous. The transformation of cities from only consumers of food to generators of agricultural products contributes to sustainability, improved health, and poverty alleviation.

- UPA assists to close the open loop system in urban areas characterized by the importation of food from rural zones and the exportation of waste to regions outside the city or town.
- Wastewater and organic solid waste can be transformed into resources for growing agriculture products: the former can be used for irrigation, the latter as fertilizer.
- Vacant urban areas can be used for agriculture production.
- Other natural resources can be conserved. The use of wastewater for irrigation increases the availability of freshwater for drinking and household consumption.
- UPA can help to preserve bioregional ecologies from being transformed into cropland.
- Urban agriculture saves energy (e.g. energy consumed in transporting food from rural to urban areas).
- Local production of food also allows savings in transportation costs, storage, and in product loss, what results in food cost reduction.
- UPA improves the quality of the urban environment through greening and thus, a reduction in pollution.

Health Improvement

- Urban agriculture also makes of the city a healthier place to live by improving the quality of the environment.
- UPA is a very efficient tool to fight against hunger and malnutrition since it facilitates the access to food by an impoverished sector of the urban population.

Poverty Alleviation: It is known that a large part of the people involved in urban agriculture is the urban poor. In developing countries, the majority of urban agricultural production is for self-consumption, with surpluses being sold in the market. According to the FAO (Food and Agriculture Organization of the United Nations), urban poor consumers spend between 60 and 80 percent of their income on food, making them very vulnerable to higher food prices.

- UPA provides food and creates savings in household expenditure on consumables, thus increasing the amount of income allocated to other uses.
- UPA surpluses can be sold in local markets, generating more income for the urban poor.

Community centers and gardens educate the community to see agriculture as an integral part of urban life. The Florida House Institute for Sustainable Development in Sarasota, Florida, serves as a public community and education center in which innovators with sustainable, energy-saving ideas can implement and test them.

Community centers like Florida House provide urban areas with a central location to learn about urban agriculture and to begin to integrate agriculture with the urban lifestyle. Other examples of community centers are Greensgrow Farm in Philadelphia and Northey Street City Farm in Brisbane, Australia. Greensgrow uses an abandoned site as an urban farm to teach the community how food is grown and how to grow their own food.

Northey Street City Farm hosts weekly community activities to educate and involve local residents in agricultural practices. Urban farms also are a proven effective educational tool to teach kids about healthy eating and meaningful physical activity. An example of educational urban agriculture is Full Circle Farm, an 11-acre farm located on a middle school campus in the heart of Silicon Valley. Like other educational agriculture centers, Full Circle Farm's acreage is used as a "living campus" where students get real-world, hands-on agriculture experiences that cultivate both healthy habits and environmental leadership.

Difficulties of Urban Agriculture

- The use of waste water for irrigation without careful treatment and monitoring can result in the spread of diseases among the population.
- Cultivation on contaminated land represents also a health hazard for the consumers.
- The practice of cultivating along roadsides facilitates the distribution of products to local markets, but it is at the same time a risky practice since it exposes food to car pollution.
- Agriculture and urbanization are considered to be incompatible activities, competing for the access and use of limited land. In reality, in urban areas there is important available space for agriculture use such as public and private vacant lots, and areas not suited for built-up uses (steep slopes and flood plains).
- Legal restrictions and economic impediments to accessing land and resources (such as reasonably priced water) are among the most common problems confronted by urban agriculture.
- Lack of security of tenure also acts as a preventive for farming due to the uncertainty in the use length of the land.

- Urban agriculture has been criticized by those who believe that industrial farm production can produce food at larger volumes more efficiently.
- A major argument is whether urban farming alone-that is farming very intensively on small land areas-could replace land extensive production in rural areas which produce the bulk of our food products. Yet, hunger persists in both urban and rural areas, despite a subsidized industrial agriculture. The degree to which urban agriculture can address these food needs systemically is undetermined, though there are indications in some communities it is an important source of food.
- Other opponents argue that localized food production and the introduction of common resources and common lands into the urban areas would produce a tragedy of the commons. Though, as referenced earlier, many urban farms and community gardens are managed privately or through other civil society organizations.

ORGANIC FARMING

Organic farming is a form of agriculture that relies on crop rotation, green manure, compost, biological pest control, and mechanical cultivation to maintain soil productivity and control pests, excluding or strictly limiting the use of synthetic fertilizers and synthetic pesticides, plant growth regulators, livestock feed additives, and genetically modified organisms. Since 1990 the market for organic products has grown at a rapid pace, averaging 20-25 percent per year to reach $33 billion in 2005. This demand has driven a similar increase in organically managed farmland. Approximately 306,000 square kilometres (30.6 million hectares) worldwide are now farmed organically, representing approximately 2% of total world farmland. In addition, as of 2005 organic wild products are farmed on approximately 62 million hectares. (IFOAM 2007:10). Organic agricultural methods are internationally regulated and legally enforced by many nations, based in large part on the standards set by the International Federation of Organic Agriculture Movements (IFOAM), an international umbrella organization for organic organizations established in 1972. IFOAM defines the overarching goal of organic farming as follows: "Organic agriculture is a production system that sustains the health of soils, ecosystems and people. It relies on ecological processes, biodiversity and cycles adapted to local conditions, rather than the use of inputs with adverse effects. Organic agriculture combines tradition, innovation and science to benefit the shared environment and promote fair relationships and a good quality of life for all involved.." —International Federation of Organic Agriculture Movements .

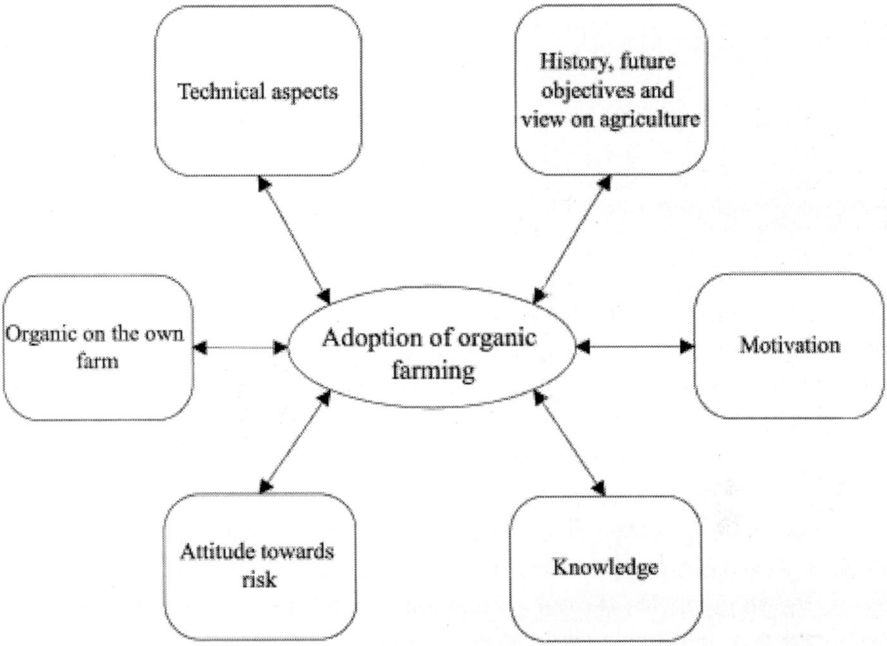

Figure: Aspects affecting decision to convert to organic farming

The organic movement began in the early 1930s and early 1940s as a reaction to agriculture's growing reliance on synthetic fertilizers. Artificial fertilizers had been created during the 18th century, initially with superphosphates and then ammonia derived fertilizers mass-produced using the Haber-Bosch process developed during World War I. These early fertilizers were cheap, powerful, and easy to transport in bulk. Similar advances occurred in chemical pesticides in the 1940s, leading to the decade being referred to as the 'pesticide era'.

Sir Albert Howard is widely considered to be the father of organic farming. Further work was done by J.I. Rodale in the United States, Lady Eve Balfour in the United Kingdom, and many others across the world.

As a percentage of total agricultural output, organic farming has remained tiny since its beginning. As environmental awareness and concern increased amongst the general population, the originally supply-driven movement became demand-driven. Premium prices from consumers and in some cases government subsidies attracted many farmers into converting. In the developing world, many farmers farm according to traditional methods which are comparable to organic farming but are not certified. In other cases, farmers in the developing world have converted out of necessity. As a proportion of total global agricultural output, organic output remains small, but it has been growing rapidly in many countries, notably in Europe.

Standards

Standards regulate production methods and in some cases final output for organic agriculture.

Standards may be voluntary or legislated. As early as the 1970s organic producers could be voluntarily certified by private associations. In the 1980s, governments began to produce organic production guidelines. Beginning in the 1990s, a trend toward legislation of standards began, most notably with the 1991 EU-Eco-regulation developed for European Union, which set standards for 12 countries, and a 1993 UK program. The EU's program was followed by a Japan program in 2001, and in 2002 the United States created the National Organic Program (NOP). As of 2007 over 60 countries have regulations on organic farming (IFOAM 2007:11). In 2005 IFOAM created the Principles of Organic Agriculture, an international guideline for certification criteria. Typically the agencies do not certify individual farms, but rather accredit certification groups. Materials used in organic production and foods are tested independently by the Organic Materials Review Institute.

Methods

"An organic farm, properly speaking, is not one that uses certain methods and substances and avoids others; it is a farm whose structure is formed in imitation of the structure of a natural system that has the integrity, the independence and the benign dependence of an organism" —Wendell Berry, "The Gift of Good Land"

The term holistic is often used to describe organic farming. Enhancing soil health is the cornerstone of organic farming. A variety of methods are employed, including crop rotation, green manure, cover cropping, application of compost, and mulching. Organic farmers also use certain processed fertilizers such as seed meal, and various mineral powders such as rock phosphate and greensand, a naturally occurring form of potash. These methods help to control erosion, promote biodiversity, and enhance the health of the soil.

Pest control targets animal pests (including insects), fungi,weeds and disease. Organic pest control involves the cumulative effect of many techniques, including, allowing for an acceptable level of pest damage, encouraging or even introducing beneficial organisms, careful crop selection and crop rotation, and mechanical controls such as row covers and traps. These techniques generally provide benefits in addition to pest control—soil protection and improvement, fertilization, pollination, water conservation, season extension, etc.—and these benefits are both complementary and cumulative in overall effect on farm health. Effective organic pest control requires a thorough understanding of pest life cycles and interactions.

Most organic farms use less pesticides than most conventional farms. The main three used are Bt (a bacterial toxin), pyrethrum and rotenone. Surveys have found that fewer than 10% of organic farmers use these pesticides regularly; one survey found that only 5.3% of vegetable growers in California use rotenone while 1.7% use pyrethrum (Lotter 2003:26).

Weeds are controlled mechanically, thermically and through the use of covercrops and mulches. The traditional method is to remove weeds by hand, which is still practiced in developing countries by small scale farmers. However, this has proven too costly in developed countries where labor is more expensive. One recent innovation in rice farming is to introduce ducks and fish to wet paddy fields, which eat both weeds and insects.

Composting

Under USDA organic standards, manure must be composted and allowed to reach a sterilizing temperature. If raw animal manure is used, 120 days must pass before the crop is harvested.

Economics

The economics of organic farming, a subfield of agricultural economics, encompasses the entire process and effects of organic farming in terms of human society, including social costs, opportunity costs, unintended consequences, information asymmetries, and economies of scale. Although the scope of economics is broad, agricultural economics tends to focus on maximizing yields and efficiency at the farm level.

Mainstream economics takes an anthropocentric approach to the value of the natural world: biodiversity, for example, is considered beneficial only to the extent that it is valued by people and increases profits. Some governments such as the European Union subsidize organic farming, in large part because these countries believe in the external benefits of reduced water use, reduced water contamination by pesticides and nutrients of organic farming, reduced soil erosion, reduced carbon emissions, increased biodiversity, and assorted other benefits.

Organic farming is labor and knowledge-intensive whereas conventional farming is capital-intensive, requiring more energy and manufactured inputs. Organic farmers in California have cited marketing as their greatest obstacle.

Geographic Producer Distribution

The markets for organic products are strongest in North America and Europe, which as of 2001 are estimated to have $6 and $8 billion respectively of the $20 billion market (2003:6). However, as of 2007 organic farmland is distributed across the globe. Australasia has 39% of the total organic farmland with Australia's 11.8 million hectares, but 97 percent of this land is sprawling rangeland (2007:35), which results in total sales of approximately 5% of US sales (2003:7). Europe has 23 percent of total organic farmland (6.9 million hectares), followed by Latin America with 19 percent (5.8 million hectares). Asia has 9.5 percent while North America has 7.2 percent. Africa has a mere 3 percent.

Besides Australia, the countries with the most organic area are Argentina (3.1 million hectares), China (2.3 million hectares), and the United States (1.6 million hectares). Much of Argentina's organic farmland is pasture, like that of Australia (2007:42). Italy, Spain, Germany, Brazil, Uruguay, and the UK follow the United States by the amount of land managed organically (2007:26).

Growth

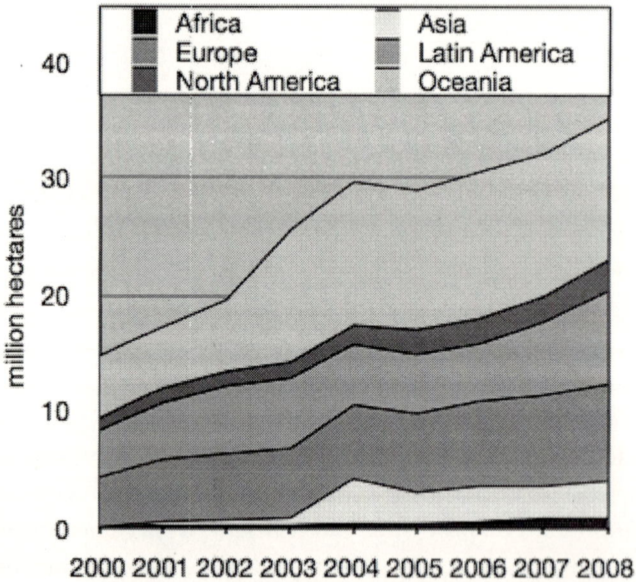

Figure: Organic farmland by world region (2000-2008)

As of 2001, the estimated total market value of certified organic products was estimated to be $20 billion. By 2002 this was $23 billion and by 2005 $33 billion, with Organic Monitor projecting sales of $40 billion in 2006 (IFOAM 2007:11). The change from 2001 to 2005 represents a compound growth of 10.6 percent.

In recent years both Europe and North America have experienced strong growth in organic farmland. Each added half a million hectares from 2004 to 2007 – for the US this is a 29 percent change (IFOAM 2007:11,27). However, this growth has occurred under different conditions. While the European Union has shifted agricultural subsidies to organic farmers in recognition of its environmental benefits, the United States has taken a free market approach. As a result, as of 2001 3 percent of European farmland was organically managed compared to just 0.3 percent of United States farmland (Lotter 2003:7). By 2005 Europe's organic land was 3.9 percent while the United States' had risen to 0.6 percent (IFOAM 2007:14-15).

IFOAM's The World of Organic Agriculture: Statistics and Emerging Trends 2007 lists the countries which added the most hectares and had the highest percentage growth in 2007 (IFOAM 2007:27-28). Among these, China is listed third in adding the most hectares behind the United States and Argentina. China jumped from approximately 300,000 hectares of organic land in 2005 to approximately 3.5 million hectares in 2006 – an increase of over a thousand percent. This rise can be attributed to the certification of China's Organic Food Development Center in 2002 by IFOAM. The end of 2005 marks the end of the three-year transition period begun in 2002..

Productivity and Profitability

A 2006 study suggests that converted organic farms have lower pre-harvest yields than their conventional counterparts in developed countries (92%) and that organic farms have higher pre-harvest yields than their low-intensity counterparts in developing countries (132%). The researcher attributes this to a relative lack of expensive fertilizers and pesticides in the developing world compared to the intensive, subsidy-driven farming of the developed world. Nonetheless, the researcher purposely avoids making the claim that organic methods routinely outperform green-revolution (conventional) methods. This study incorporated a 1990 review of 205 crop comparisons which found that organic crops had 91% of conventional yields. A major US survey published in 2001, analyzed results from 150 growing seasons for various crops and concluded that organic yields were 95-100% of conventional yields.

Lotter (2003:10) reports that repeated studies have found that organic farms withstand severe weather conditions better than conventional farms, sometimes yielding 70-90% more than conventional farms during droughts. A 22-year farm trial study by Cornell University published in 2005 concluded that organic farming produces the same corn and soybean yields as conventional methods over the long-term averages, but consumed less energy and used zero pesticides.

The results were attributed to lower yields in general but higher yields during drought years. A study of 1,804 organic farms in Central American hit by Hurricane Mitch in 1998 found that the organic farms sustained the damage much better, retaining 20 to 40% more topsoil and smaller economic losses at highly significant levels than their neighbors.

On the other hand, a prominent 21-year Swiss study found an average of 20% lower organic yields over conventional, along with 50% lower expenditure on fertilizer and energy, and 97% less pesticides.

A long-term study by U.S Department of Agriculture Agricultural Research Service (ARS) scientists concluded that, contrary to widespread belief, organic farming can build up soil organic matter better than conventional no-till farming, which suggests long-term yield benefits from organic farming. An 18-year study of organic methods on nutrient-depleted soil concluded that conventional methods were superior for soil fertility and yield in a cold-temperate climate, arguing that much of the benefits from organic farming are derived from imported materials which could not be regarded as "self-sustaining".

While organic farms have lower yields, organic methods require no synthetic fertilizer and pesticides. The decreased cost on those inputs, along with the premiums which consumers pay for organic produce, create higher profits for organic farmers. Organic farms have been consistently found to be as or more profitable than conventional farms with premiums included, but without premiums profitability is mixed (Lotter 2003:11). Welsh (1999) reports that organic farmers are more profitable in the drier states of the United States, likely due to their superior drought performance.

In 2008 the UN Environmental Programme (UNEP) and UN Conference on Trade and Development (UNCTAD) issued a report which stated that "organic agriculture can be more conducive to food security in Africa than most conventional production systems, and that it is more likely to be sustainable in the long-term". The report assessed 114 projects in 24 African countries, finding that "yields had more than doubled where organic, or near-organic practices had been used" and that soil fertility and drought resistance improved.

Macroeconomic Impact

Organic methods often require more labor, providing rural jobs but increasing costs to urban consumers.

Motivations

Agriculture in general imposes external costs upon society through pesticides, nutrient runoff, excessive water usage, and assorted other problems. As organic methods minimize some of these factors, organic farming is believed to impose fewer external costs upon society. A 2000 assessment of agriculture in the UK determined total external costs costs for 1996 of 2343 million British pounds or 208 pounds per hectare. A 2005 analysis of these costs in the USA concluded that cropland imposes approximately 5 to 16 billion dollars ($30 to $96 per hectare), while livestock production imposes 714 million dollars. Both studies concluded that more should be done to internalize external costs, and neither included subsidies in their analysis, but noted that subsidies also influence the cost of agriculture to society. Both focused on purely fiscal impacts. The 2000 review included reported pesticide poisonings but did not include speculative chronic effects of pesticides, and the 2004 review relied on a 1992 estimate of the total impact of pesticides.

Pesticides

Some pesticides may damage human health and the environment, and most organic farms use less pesticides than conventional farms. The main three pesticides used in organic farming are Bt (a bacterial toxin), pyrethrum, rotenone, copper and sulphur. Surveys have found that fewer than 10% of organic farmers use these pesticides regularly; one survey found that only 5.3% of vegetable growers in California use rotenone while 1.7% use pyrethrum (Lotter 2003:26). Reduction and elimination of chemical pesticide use is technologically challenging. Few organic farms manage to eliminate the use of pesticides entirely; organic pesticides are often used to compliment other pest control strategies.

Pesticide runoff is one of the most significant effects of pesticide use. The USDA Natural Resources Conservation Service tracks the environmental risk posed by pesticide water contamination from farms, and its conclusion has been that "the Nation's pesticide policies during the last twenty six years have succeeded in reducing overall environmental risk, in spite of slight increases in area planted and weight of pesticides applied. Nevertheless, there are still areas of the country where there is no evidence of progress, and areas where risk levels for protection of drinking water, fish, algae and crustaceans remain high". Pest resistant genetically modified crops have been proposed as an alternative to pesticide use, however concerns over the safety and the long term benefits of genetically modified food, result in the genetic modification being widely opposed in the organic farming movement.

Genetically Modified Organisms

A key characteristic of organic farming is rejection of genetically engineered products, including plants and animals. On October 19, 1998, participants at IFOAM's 12th Scientific Conference of IFOAM) issued the Mar del Plata Declaration, where more than 600 delegates from over 60 countries voted unanimously to exclude the use of genetically modified organisms in food production and agriculture.

From this point, it became widely recognized that GMOs are categorically excluded from organic farming. Although opposition to the use of any transgenic technologies in organic farming is strong, agricultural researchers Luis Herrera-Estrella & Ariel Alvarez-Morales continue to advocate integration of transgenic technologies into organic farming as the optimal means to sustainable agriculture, particularly in the developing world. Similarly, some organic farmers question the rationale behind the ban on the use of genetically engineered seed because they see it a biological technology consistent with organic principles

Although GMOs are excluded from use in organic farming, there is concern that the pollen from genetically modified crops is increasingly contaminating organic and heirloom genetics making it difficult, if not impossible, to keep these genetics from entering the organic food supply. International trade restrictions limit the availability GMOs to certain countries.

The actual dangers that genetic modification could pose to the environment or, supposedly, individual health, are hotly contended.

Food Quality

Organic food is widely believed by the lay public to be healthier, although the research is controversial. Two studies have found that children fed organic diets experienced significantly lower organophosphorus pesticide exposure than children fed conventional diets. Although the researchers did not collect health *outcome* data in this study, they concluded "it is intuitive to assume that children whose diets consist of organic food items would have a lower probability of neurologic health risks". A 2007 study found that consumption of organic milk is associated with a decrease in risk for eczema, although no comparable benefit was found for organic fruits, vegetables, or meat.

Extensive scientific research is being carried out in Switzerland at over 200 farms to determine differences in the quality of organic food products compared to conventional in addition to other tests. The FiBL Institute has been investigating the differences at over 200 farms. It states that "organic products stand out as having higher levels of secondary plant compounds and vitamin C. In the case of milk and meat, the fatty acid profile is often better from a nutritional point of view. As far as carbohydrates and minerals, organic products are no different from conventional products. However, in regard to undesirables such as nitrate and pesticide residues, organic products have a clear advantage. A £12m EU-funded investigation into the difference between organic and ordinary farming published in 2007 found that organic foods have more nutritional value. A recent study found that organically grown produce has double the flavonoids, an important antioxidant.. A 2007 study found that organically grown kiwifruit had more antioxidants than conventional kiwifruit.

Soil Conservation

In *Dirt: The Erosion of Civilizations*, geomorphologist David Montgomery outlines a coming crisis from soil erosion. Agriculture relies on roughly one meter of topsoil, and that is being depleted ten times faster than it is being replaced. No-till farming, which some claim depends upon pesticides, is regarded as one way to minimize erosion. However, a recent study by the USDA's Agricultural Research Service has found that organic farming is even better at building up the soil than no-till. Cite error: Closing </ref> missing for <ref> tag According to the Rodale Institute, which has been comparing organic agricultural systems and conventional systems since 1981, organic agriculture also can be used to mitigate global warming by decreasing fossil fuel emissions and sequestering carbon in the soil. The elimination of synthetic nitrogen in organic systems decreases fossil fuel consumption by 33 percent (LaSalle)and carbon sequestration takes CO2 out of the atmosphere by putting it in the soil in the form of organic matter which is often lost in conventionally managed soils. Carbon sequestration occurs at especially high levels in organic no-till managed soil according to the Rodale Institute.

Nutrient Leaching

Excess nutrients in lakes, rivers, and groundwater can cause algal blooms, eutrophication, and subsequent dead zones. In addition, nitrates are harmful to aquatic organisms by themselves. The main contributor to this pollution is nitrate fertilizers whose use is expected to "double or almost triple by 2050". Researchers at the United States National Academy of Sciences found that organically fertilizing fields "significantly [reduces] harmful nitrate leaching" over conventionally fertilized fields: "annual nitrate leaching was 4.4-5.6 times higher in conventional plots than organic plots". Scientists believe that the large dead zone in the Gulf of Mexico is caused in large part by agricultural pollution: a combination of fertilizer runoff and livestock manure runoff. A study by the United States Geological Survey (USGS) found that over half of the nitrogen released into the Gulf comes from agriculture. The economic cost of this for fishermen may be large, as they must travel far from the coast to find fish.

At the 2000 IFOAM Conference, researchers presented a study of nitrogen leaching into the Danube River. They found that nitrogen runoff was substantially lower among organic farms and suggested that the external cost could be internalized by charging 1 euro per kg of nitrogen released. A 2005 study found a strong link between agricultural runoff and algae blooms in California.

Biodiversity

A wide range of organisms benefit from organic farming, but it is unclear whether organic methods confer greater benefits than integrated agri-environmental conventional programs. Nearly all non-crop, naturally-occurring species observed in comparative farm land practice studies show a preference in organic farming both by population and richness. Spanning all associated species, there is an average of 30% more on organic farms versus conventional farming methods. Birds, butterflies, soil microbes, beetles, earthworms, spiders, vegetation, and mammals are particularly affected.

Organic crops use little or no herbicides and pesticides and thus biodiversity fitness and population density benefit. Many weed species attract beneficial insects that improve soil qualities and forage on weed pests. Soil-bound organisms often benefit because of increased bacteria populations due to natural fertilizer spread such as manure, while experiencing reduced intake of herbicides and pesticides commonly associated with conventional farming methods. Increased biodiversity, especially from soil microbes such as mycorhizzae, have been proposed as an explanation for the high yields experienced by some organic plots, especially in light of the differences seen in a 21-year comparison of organic and control fields.

The level of biodiversity that can be yielded from organic farming provides a natural capital to humans. Species found in most organic farms provides a means of agricultural sustainability by reducing amount of human input (e.g. fertilizers, pesticides). Farmers that produce with organic methods reduce risk of poor yields by promoting biodiversity. Common game birds such as the ring-necked pheasant and the northern bobwhite often reside in agriculture landscapes, and are a natural capital yielded from high demands of recreational hunting. Because bird species richness and population are typically higher on organic farm systems, promoting biodiversity can be seen as logical and economical. Biological research on soil and soil organisms has proven beneficial to the system of organic farming. Varieties of bacteria and fungi break down chemicals, plant matter and animal waste into productive soil nutrients. In turn, the producer benefits by healthier yields and more arable soil for future crops.

Furthermore, a 21-year study was conducted testing the effects of organic soil matter and its relationship to soil quality and yield. Controls included actively managed soil with varying levels of manure, compared to a plot with no manure input. After the study commenced, there was significantly lower yields on the control plot when compared to the fields with manure. The concluded reason was an increased soil microbe community in the manure fields, providing a healthier, more arable soil system.

Distributors

In the United States, 75% of organic farms are smaller than 2.5 hectares and in California 2% of the farms account for over half of the sales (Lotter 2003:4). Groups of small farms join together in cooperatives such as Organic Valley, Inc. to market their goods more effectively.

Over the past twenty years, however, most of these cooperative distributors have merged or been bought out. Rural sociologist Philip H. Howard has researched the structure and transformation of the organic industry in the United States. He claims that in 1982 there were 28 consumer cooperative distributors but as of 2007 there are only 3, and he has created a graphic displaying the consolidation. His research shows that most of these small cooperatives have been absorbed into large multinational corporations such as General Mills, Heinz, ConAgra, Kellog, and assorted other brands. This consolidation has raised concerns among consumers and journalists of potential fraud and degradation in standards. Most of these large corporations sell their organic products through subsidiaries, allowing them to keep their names off the labels.

Sales and Marketing

Organic farmers report that marketing and distribution are difficult obstacles. Most of organic sales are concentrated in developed nations. These products are what economists call credence goods in that they rely on uncertain certification. As food prices rise, organic products may experience falling demand. A 2008 survey by WSL Strategic Retail found that interest in organic products had dropped since 2006, and that 42% of Americans polled don't trust organic produce. and The Hartman Group reports that 69% of Americans claim to occasionally buy organic products, down from 73% in 2005. The Hartman Group says that people may be substituting local produce for organic produce.

Farmers' Markets

Price premiums are important for the profitability of small organic farmers, and so many sell directly to consumers in farmers' markets. In the United States the number of farmers' markets has grown from 1,755 in 1994 to 4,385 in 2006.

Controversy

A number of critics contest the notion that organic agricultural systems are more friendly to the environment and more sustainable than high-yielding farming systems. Among these critics are Norman Borlaug, father of the "green revolution," and winner of the Nobel Peace Prize, who asserts that organic farming practices can at most feed 4 billion people, after expanding cropland dramatically and destroying ecosystems in the process and Prof A. Trewavas.

The debate has been summarized in an exchange between Trewavas and Lord P. Melchett, and published by a major supermarket, concerned about examining the issues.

One study from the Danish Environmental Protection Agency found that, area-for-area, organic farms of potatoes, sugar beet and seed grass produce as little as half the output of conventional farming.

In 2008 a study from UN Environmental Programme concluded that organic methods greatly increase yields in Africa and a review of over two hundred crop comparisons argued that organic farming could produce enough food per capita to sustain the current human population; the difference in yields between organic and non-organic methods were small, with non-organic methods resulting in slightly higher yields in developed areas and organic methods resulting in slightly higher yields in developing areas.

That analysis has been severely criticised by Alex Avery, who contends that the review claimed many non-organic studies to be organic, misreported organic yields, made false comparisons between yields of organic and non-organic studies which were not comparable, counted high organic yields several times by citing different papers which referenced the same data, and gave equal weight to studies from sources which were not impartial and rigorous university studies.

Urs Niggli, director of the FiBL Institute contents that the wave of newspaper articles like 'Organic food exposed' or 'The hypocrisy of organic farmers' are a part of a global campaign against organic farming that take their arguments mostly from the book 'The truth about organic farming', by Alex Avery of the Hudson Institute.

In 1998, Dennis Avery of the Hudson Institute claimed the risk of E. coli infection was eight times higher when eating organic food rather than non-organic food, using the Center for Disease Control (CDC) as a source. When the CDC was contacted, it stated that there was no evidence for the claim. The *New York Times* commented on Avery's attacks: "The attack on organic food by a well-financed research organization suggests that, though organic food accounts for only 1 percent of food sales in the United States, the conventional food industry is worried."

Organic agriculture can reduce the level of negative externalities from (conventional) agriculture. Whether this is seen as private or public benefits depends upon the initial specification of property rights. However, it is clear that agriculture has been undervalued and underestimated as a means to combat global climate change. Soil carbon data recorded by The Rodale Institute show that regenerative organic agricultural practices are among the most effective strategies for mitigating CO_2 emissions.

Capacity Building

Organic agriculture can contribute to meaningful socio-economic and ecologically sustainable development, especially in poorer countries. On one hand, this is due to the application of organic principles, which means efficient management of local resources (e.g. local seed varieties, manure, etc.) and therefore cost-effectiveness. On the other hand, the market for organic products – at local and international level – has tremendous growth prospects and offers creative producers and exporters in the South excellent opportunities to improve their income and living conditions.

Organic Agriculture is a very knowledge intensive production system. Therefore capacity building efforts play a central role in this regard. There are many efforts all around the world regarding the development of training material and the organization of training courses related to Organic Agriculture. Big parts of existing knowledge is still scattered and not easy accessible. Especially in Developing Countries this situation remains an important constraint for the growth of the organic sector.

For that reason, the International Federation of Organic Agriculture Movements created an Internet Training Platform whose objective is to become the global reference point for Organic Agriculture training through free access to high quality training materials and training programs on Organic Agriculture. In November 2007, the Training Platform hosted more than 170 free manuals and 75 training opportunities.

AGRICULTURAL PLANNING AND DEVELOPMENT

India is a vast country with a variety of landforms, climate, geology, physiography, and vegetation India is endowed with regional diversities for its uneven "economic and agricultural" development, on account of (i) Agro-climatic environments (15 Zones/127 regions), (ii) Agro-ecological regions (20) and 60 sub-regions, (iii) Agro-Edephic regions, (iv) Terrain mapping sub-units, (v) Natural resources endowments (geology, geomorphology, soil, ground water, surface water, & infrastructure), (vi) Human resources (Population density), (vii) Level of investments in rural infrastructure, and (viii) Level of investment in technology and its adoption. India has a total geographical area (TGA) of 329 Million Hectares (MH) out of which, about 265 MH represent varying degrees of potential for biological production. [Dhuruva89] report reveals that more than 50% of TGA is threatened by various types of land degradation, such as soil erosion, gully & ravine formation, salinity, water logging, shifting cultivation, etc. Development of irrigation potential is considered as the key factor in the sustenance of "Green Revolution". Despite 50 years of development planning, rainfed agriculture is the largest and the most important sector of crop production in India.

Soil resources are the most precious non-renewable vital resources for growing food, fibre, and fuel wood to meet the human needs. Management of Soil Resources is essential for both the continued agricultural productivity and protection of environment. By considering various factors like population growth rate, diminishing per capita of land and water resources, and increasing land degradation problems, it is estimated that India will be required to produce an additional 5-6 million tons of food grains annually in 21st Century. This will lead to tremendous pressure on soil resources along with competitive demand for it from industrialization and urbanization. However the capacity of soil to produce is limited and its limits to production are set by its inherent characteristics, agro-ecological settings, and its use and management.

Forests are an important natural resources of India, having a moderating influence against floods and also protecting the soil against erosion. About 95% of the forests in India is owned by States and the total area under forests is about 22% of the total geographical area.

Development of livestock has been envisaged as an integral part of sound system of diversified agriculture. In animal production, the major aim is for raising ecologically adapted animals and efficient utilization of locally available feed resource. Dairy development is intimately linked with cattle population, breed improvement, cattle health and disease management, and fodder development, etc. Animal Husbandry in India is essentially a endeavour of millions of small holders (Resource-Poor-Farmers) who rear animals on "crop residues" and "common property resources" without generally allowing them to compete with man for food grains.

The small holders produces milk, meat, wool, etc., for the community, with virtually no capital, resource, training and at a cost that no modern technology in the world had ever produced. Food and Fodder Resources will be crucial to the future development of "livestock resources" in the Country. There is very little scope for increasing the area under fodder production, keeping in view the priority for food grains, pulses and oil seeds. Development of Fodder Resources is basically an activity based on a multi-disciplinary approach involving the areas of agriculture, animal husbandry, environment & forests, revenue, rural development, and wasteland development. Water Resources of India contain diverse group of flora and fauna. Agriculture is the greatest user of Water accounting for about 80% of all consumption. Animal Husbandry and Fisheries require abundant water. Development of Water Resources, since Independence, has been undertaken for specific purposes like irrigation, flood control, hydro-power generation, drinking water supply, industrial and various miscellaneous uses. Minor irrigation projects have both surface and ground water as their source, while major and medium projects mostly exploit surface water resources. The break up of the ultimate irrigation potential under the above three categories is,

- 58 M.Ha by major and medium irrigation projects,
- 17 M.Ha by minor surface water schemes, and
- 64 M.Ha by minor ground water schemes.

Fisheries Resources of India are either inland or marine. The principal rivers and the tributaries, canals, ponds, lakes, reservoirs comprise inland fisheries. The river extend about 27,200 kms, and other subsidiary water channel comprise about 112,000 kms. Marine resources comprises of about 2 Million sq.kms of EEZ for deep sea fishing, and 7,250 kms of coastline. With the diverse fish fauna, the development objectives are to judiciously & optimally utilize the resources for [NBFGR2K]:-

- enhancing production and productivity of fishermen, fish farmers and fishing industry;
- increasing fish production and thereby, raising nutritional standard of people;
- earning of foreign exchange from export of marine products;
- improving Socioeconomic conditions of traditional fishermen;
- generating employment for coastal and rural poor; and
- conservation of depleting species of fish.

Good infrastructure helps in raising productivity and lowering the unit cost in the production activities of the economy. "Agricultural Infrastructure" refers to "Rural Infrastructure" whereas "Industrial Infrastructure" refers to "Urban Infrastructure". Agricultural development requires (i) agricultural research and extension, (ii) rural financial institution, (iii) irrigation and drainage, (iv) agricultural inputs (fertilizers, seeds, credits), and (v) marketing and storage facilities.

Agriculture Credit is a crucial input for increasing agricultural production and productivity. Institutional finance for Agricultural credit is disbursed mainly by Commercial banks, Regional Rural Banks, Land Development Banks, and Cooperative banks. Share of commercial banks in total institutional credit to agriculture is about 48%, that of Cooperative banks is about 46%, and Regional Rural Banks account for 6% only. Short-term Credit accounts for 2/3rd of the total institutional lending to the Agriculture.

Drought has multiplier effect on agricultural production during the subsequent year also, due to (i) non-availability of quality seeds for sowing of crops, (ii) inadequate draught power for carrying out agricultural operations as a result of either distress sale of cattle or loss of life, (iii) reduced use of fertilizers as the investment capacity of the farmers decline, (iv) non-availability of raw materials in agro-based industries, and (v) deforestation to meet the energy needs in domestic sector as agricultural waste may not be available in required quantity.

The Central Ministry of Agriculture (MOA) is responsible for implementation and formulation of national policies and programs to achieve agricultural growth through optimum utilization of the land resources, water, soil, plant, fisheries, & livestock resources. Government of India implements the following agricultural related Schemes (whether Watershed based or Agro-climatic region based) in the country, which deal agricultural resources information for Planning and Development:-

- Agro-climatic Regional Planning (ACRP) Project
- Agro-Ecological Mapping Project of the National Bureau of Soil Survey & Land Use Planning (NBSS&LUP)
- All India Soil and Land Use Survey (AISLUS)
- Early Warning System of Agricultural Situation in India
- Forecasting of Agricultural output using Space, Agro-meteorology and Land based observations (FASAL) Project
- Land Records Computerisation Project
- National Agricultural Research Project (NARP)
- National Agricultural Technology Project (NATP) to strengthen research-extension-farmer (r-e-f) linkage
- National Watershed Development Program for Rain-fed Areas (NWDPRA)
- Soil and Water Conservation Programs
- Drought Prone Area Development programme
- Desert Development Programme
- National Wastelands Development programme
- Integrated Mission on Sustainable Development (IMSD) Programme.

URBAN AGRICULTURE INTO LAND MANAGEMENT

During the research process, the topic of inserting Urban Agriculture into the land management system was addressed directly and indirectly in the research application at the regional level and also in urban agriculture as an evolving activity in the city. This process proved the need to develop a broader proposal taking Urban Agriculture to a higher level of institutionalization and defining its place within the city's management and control structures. A workshop was carried out for this purpose with the following participants: the research team, Physical Planning Directorate representatives, the Capital City Integral Development Group, the Faculty of Geography, Habitat Cuba, the Urban Agriculture Directorate of the city, a producer of the Camilo Cienfuegos region, the Faculty of Architecture's Urban Studies Centre of the Instituto Superior Politecnico Jose Antonio Echevarria, and other experts. Given the importance of the series of definitions that had to be established before arriving at the proposal for how urban agriculture could be inserted into the land management system, a summary is provided below:

Why Insert Urban Agriculture into the Land Management System?

- o Urban agriculture is a significant activity with a strong presence in the use of the land, as well as production and consumption activities in almost all the municipalities of Havana.
- o Urban agriculture was created by the government and accepted enthusiastically by different sectors of the population.
- o The economic and environmental results of urban agricultural activities in Cuba have been positive.
- o It is important to continue with urban agricultural activities beyond the economic crisis, natural disasters or any other specific situations.
- o Cities have new needs, and it is important to integrate urban agriculture as one of them.
- o Urban agriculture should be placed at a higher institutionalization level: proper recognition of its functions and requirements for a legal framework.
- o Urban agriculture should be developed, managed and designed from a land-management point of view.
- o Urban agriculture should be linked with other urban components.
- o Urban management activities should be linked with other related institutions.
- o Urban agriculture activities should be controlled.
- o The city will always have vacant spaces.
- o Urban agriculture contributes to the achievement of other urban planning objectives (quality of life, value of urban soils).
- o Urban agriculture should be addressed as a formal activity of the city.
- o Urban agriculture would be better positioned for reaching its full potential in relation to the various urban sectors (space in urban projects, commercial relation with the tourism sector).

There are powerful and sufficient reasons in support of inserting urban agriculture into the land management system.

What would be the basis for stating that urban agriculture should or must be regarded as an urban function?

- o Urban agriculture is behaving, in practice, similarly to other urban functions in the use of spaces and the actors involved.
- o Urban agriculture is a productive activity that uses space in the city and generates output. It interrelates with other spaces and functions

in the city (housing, people, markets), and that is why it should be regarded as a permanent urban function.

- o Urban agriculture improves the quality of life by guaranteeing food security, enhancing the image of a city that is closer to nature, improving the health conditions of the urban environment, and including the most vulnerable sectors in Cuban society, such as elderly people.
- o The resources required by urban agriculture have an important impact on the use of urban waste.
- o Urban agriculture, as an urban function, can be developed with more objectivity, establishing criteria, policies and strategies for reaching a level of development in accordance with the main characteristics of the province.

Urban agriculture should be considered as a permanent function. The recognized use of the space in the medium and long terms would eliminate the possibility of considering urban agriculture a temporary activity.

As an urban function, where does the development and enhancement of urban agriculture begin?

- o As a government activity, allowing the state to generate investments and to play the role of facilitator and of governing, monitoring and regulating body of urban agriculture activities through:
  - * Local governments' contributions and resource channelling for developing urban agriculture.
  - * Local governments acting as coordinators of other government agencies, the private sector and non-governmental institutions.
- o As an individual and/or collective activity, but always having an impact on the community, urban agriculture requires participatory planning for the following reasons:
- * The community is the main actor in urban agriculture.
- * Urban agriculture promotes exchanges, self-sufficiency and local development.
- * Urban agriculture saves the government money and resources.
- * Urban agriculture is instrumental in helping the community to discover and use other possibilities.
- * Urban agriculture generates profits at the local level.
- * Urban agriculture helps to develop and improve interpersonal relations.
- * The management of the space benefits the citizens directly.
- * Urban agriculture promotes community participation.

Urban agriculture should be treated as an activity in which the primary actors are the government and the community, both having specific roles and gaining mutual benefits. As an urban function, at what work scale should urban agriculture be undertaken and on what bases?

- o The city's diversity and the homogeneity of existing spaces should be taken into account, as determined by the potential of the land and its characteristics.
- o Ciudad de La Habana province is considered a strictly urban province, but it has a peri-urban area with characteristics similar to a rural environment. It is necessary to keep in mind that parts of Havana are developed areas, others are being developed and some are not suitable for development. It is also important to distinguish between the forms of urban agriculture in the different types of zones, from the most developed areas to the least developed spaces on the city's outskirts.
- o Urban agricultural development requires a specific design, adapted to the urban space and environment. It is necessary to seek alternative solutions to problems such as water, which is more feasible on a local scale.
- o Urban agriculture must be addressed as a permanent urban function with different treatments according to the zones, either central, intermediate or outside the city, as well as on a general and a local scale.
- o The treatment of urban agriculture as a permanent function does not exclude the use of specific spaces set aside for other purposes (this is the difference between urban agriculture and other urban functions, and the difference between urban agriculture and rural agriculture). However, the use of these spaces allocated for other purposes should occur only in specific cases and for a set period of time.

Urban agriculture should be treated within the land management system on a large and medium scale in the city, and also in the urban development design, taking into account its specific nature and its links with other urban sectors. With what urban components might urban agriculture bring about a mutually beneficial relationship? When interacting with the urban structure, urban agriculture seeks to establish various links which, if well developed and enhanced, may lead to important economic, social, environmental and land-use processes. Both the city and the urban agriculture sector gain through a mutually beneficial relationship. How can urban agriculture be included in the land management process? What elements are to be considered and how can this process be addressed?

- o It is important to visualize urban agriculture as part of the physical planning system and to find an integrating conceptual framework for a common language among the institutions directly or indirectly related to urban agriculture and to the city.
- o In this context, it is important to separate the concept of urban agricultural management. Only after knowing what we expect of urban agriculture in each place, will we be able to address the question of who will manage it and how.
- o By including urban agriculture as part of the Green System, it would be possible in the Cuban context to use its full contribution to food security and go beyond the productive dimension to become part of the urban environment, fulfilling, at the same time, the different functions required by the economy, the people and the urban structure.
- o Urban agriculture should be addressed within the land management system not only on the basis of what it is now but following the recommendations of research studies for the future, including what we have learnt from past experiences in Cuba.
- o By considering urban agriculture only as a productive activity, we run the risk of excluding all other urban agriculture potential, such as involvement in the community participatory process, environmental education, and re-use of urban waste. It is an activity generating twofold benefits – for the urban agricultural sector and for the urban environment.
- o The role of urban agriculture in the housing sector goes beyond improving the image of the areas outside the houses. It has an impact on a higher scale and occupies an important place in the local economy.
- o At the general city level, the concept of urban agriculture within the physical planning system is most valuable. At the level of project and design (neighbourhood, zone) it will be necessary to study and assess the availability of resources, infrastructure, actors involved, and other uses and linkages.
- o The Green System aims at linking zones with different functions, generating spaces with different uses; improving on problems such as air pollution, noise, access to other urban spaces, and promoting healthier production and recreational activities.

The Cuban case shows that the best way of addressing urban agriculture within the land management system is to include it in the Green System.

Urban Agriculture as a Component

The city seeks green spaces emanating vitality and beauty for enriching existing spaces and responding to the citizens' needs for a new distribution of urban spaces and better quality of life. That is why it is important to organize the different spaces by ensuring their individuality and their unique formal, dimensional and qualitative differences. The Green System is based on the principle of linkages and continuity of the green spaces to ensure uses and functions according to their locations. Green spaces link urban networks and have a very important meaning for the city's image and operations. In order to address urban agriculture as part of the Green System, it is necessary to define its roles within this system in a flexible, easily implemented work plan. As part of the Green System, urban agriculture will be able to have an integrated set of activities and reach its full potential, with the flexibility to include new elements in the system.

Figure: Despite the progress with urban agriculture

Addressing urban agriculture in this context requires a series of definitions, some of which were discussed during the workshop. The workshop evaluation also recommends further studies on the subject in order to establish a methodological basis for inserting urban agriculture into the Green System. Currently, Havana's Green System includes three subsystems that, although linked together, have their own spatial and operational requirements. The evaluation carried out showed that it was more convenient to insert urban agriculture into the existing agricultural production subsystem instead of creating a new subsystem. However, this would involve redefining the agricultural production subsystem to include the following:

- To define the concept of a green area (valid for the whole system).

- To establish its components and types of spaces.
  - o Linkages among the different components.
  - o Identification of local components and their linkages.
  - o Economic and environmental indicators.
  - o Legal framework, urban rules and regulations.
  - o Characterization of spaces where, among other issues, it would be necessary to address the question of the value of the soil beyond its agricultural production potential.
  - o Management (who manages it and how)
  - o Land ownership issues.
  - o Access to land for producers.
  - o Incentives (establishing, maintaining and developing phases).
  - o Others.

URBAN AGRICULTURE AND ORGANIC FARMING

Urban farming is generally practiced for income-earning or food-producing activities though in some communities the main impetus is recreation and relaxation. Urban agriculture contributes to food security and food safety in two ways: first, it increases the amount of food available to people living in cities, and, second, it allows fresh vegetables and fruits and meat products to be made available to urban consumers. A common and efficient form of urban agriculture is the biointensive method. Because urban agriculture promotes energy-saving local food production, urban and peri-urban agriculture are generally seen as sustainable practices.

Figure: An urban farm in Chicago

The recognition of environmental degradation within cities through the relocation of resources to serve urban populations has inspired the implementation of different schemes of urban agriculture across the developed and developing world. From historic models such as Machu Picchu to designs for new productive urban farms, the idea of locating agriculture in the city takes on many characteristics.

History of Urban Agriculture

Community wastes were used in ancient Persia to feed urban farming. In Machu Picchu water was conserved and reused as part of the stepped architecture of the city and vegetable beds were designed to gather sun in order to prolong the growing season. Victory gardens sprouted during WWI, WWII and were fruit, vegetable, and herb gardens in US, Canada, and UK. This effort was undertaken by citizens to reduce pressure on food production that was to support the war effort. Community gardening in most communities are open to the public and provide space for citizens to cultivate plants for food or recreation. A community gardening program that is well-established is Seattle's P-Patch.

RESOURCE AND ECONOMIC

The Food and Agriculture Organization of the United Nations (FAO), has defined urban agriculture as: " [A]n industry that produces, processes and markets food and fuel, largely in response to the daily demand of consumers within a town, city, or metropolis, on land and water dispersed throughout the urban and peri-urban area, applying intensive production methods, using and reusing natural resources and urban wastes to yield a diversity of crops and livestock." The definition of urban agriculture as an industry that responds to the nutritional demands of a city, from within that city, with the use and reuse of that city's resources while acknowledging economic and resource use does not reconcile aspects of regional health, food security, and application of grassroots organizations. (This definition is based on the work of Luc Mougeot of the International Development Research Centre and used in technical and training publications by UN-HABITAT's Urban Management Programme, FAO's Special Programme for Food Security, and international agricultural research centres, such as CIRAD.)

Environmental

The Council on Agriculture, Science and Technology, (CAST) is an international consortium of scientific and professional societies based in Ames

Iowa that compiles and communicates credible science-based information to policy makers, media, private sector, and the public. CAST defines urban agriculture to include aspects of environmental health, remediation, and recreation,

"Urban agriculture is a complex system encompassing a spectrum of interests, from a traditional core of activities associated with the production, processing, marketing, distribution, and consumption, to a multiplicity of other benefits and services that are less widely acknowledged and documented. These include recreation and leisure; economic vitality and business entrepreneurship, individual health and well-being; community health and well being; landscape beautification; and environmental restoration and remediation." Modern planning and design initiatives are more responsive to this model of urban agriculture because it fits within the current scope of sustainable design. The definition allows for a multitude of interpretations across cultures and time. Frequently it is tied to policy decisions to build sustainable cities.

Food Security

Access to nutritious food is another perspective in the effort to locate food and livestock production in cities. With the tremendous influx of world population to urban areas, the need for fresh and safe food is increased. Community Food Security Coalition (CFSC) defines food security as, "all persons in a community having access to culturally acceptable, nutritionally adequate food through local, non-emergency sources at all times."

The Emerging Importance of Urban Agriculture

Economic

- UPA (urban and peri-urban agriculture) expands the economic base of the city through production, processing, packaging, and marketing of consumable products. This results in an increase in entrepreneurial activities and the creation of job opportunities, as well as in food costs reduction and products of better quality.
- UPA represents an important opportunity for women to be part of the informal economy of a city. Farming and selling activities can be combined more easily with household tasks and child care.
- UPA provides employment, income, and access to food for urban populations, which together contributes to relieve chronic and emergency food insecurity. Chronic food insecurity refers to less

affordable food and growing urban poverty, while emergency food insecurity relates to breakdowns in the chain of food distribution. UPA plays an important role in making food more affordable and in providing emergency supplies of food. Research into market values for produce grown in urban gardens has attributed to a community garden plot a median yield value of between approximately $200 and $500 (US, adjusted for inflation). In a community gardening program as well-established as Seattle's P-Patches, this can account for up to 1.25 million dollars of produce cultivated annually.

Social

Social benefits that have emerged from urban agricultural practices are; better health and nutrition, increased income, employment, food security within the household, and community social life. UPA can be seen as a means of improving the livelihood of people living in and around cities. Taking part in such practices is seen mostly as informal activity, but in many cities where inadequate, unreliable, and irregular access to food is an occurring problem, urban agriculture has been a positive response to tackling food concerns. Households and small communities take advantage of vacant land and contribute not only to their household food needs but also the needs of their resident city. The CFSC states that, "Community and residential gardening, as well as small-scale farming, save household food dollars. They promote nutrition and free cash for non-garden foods and other items"

This allows families to generate larger incomes selling to local grocers or to local outdoor markets, while supplying their household with proper nutrition of fresh and nutritional produce.

Some community urban farms can be quite efficient and help women find work, who in some cases are marginalized from finding employment in the formal economy. Studies have shown that participation from women have a higher production rate, therefore producing the adequate amount for household consumption while supplying more for market sale.

Due to the fact that most UA activities are conducted on vacant municipal land, there have been rising concerns about the allocation of land and property rights. The IDRC and the FAO have published the Guidelines for Municipal Policymaking on Urban Agriculture, and are working with municipal governments to create successful policy measures that can be incorporated in urban planning. Including UA in local plans and as proper land use will continue to help impoverished communities gain a better well-being while fighting urban poverty.

Localized food production in urban and peri-urban areas contributes to local economies by creating jobs and producing valuable products. Some researchers indicate that unemployed populations in large cities and suburban towns would decrease if put to work by local food movements. Schools have foreseen the asset of local food production and are beginning to incorporate agricultural sections in their curricula and present it as a career opportunity. Urban agricultural projects are beginning to open a new labor market in areas that have been negatively affected by industrial outsourcing of jobs.

Quality of Food

Although the taste of locally grown food is subjective, many participants in the urban agriculture movement report that they prefer the taste of local agricultural products, or organic food, to that of industrial food production. Also, urban agriculture supports a more sustainable production of the food that tries to decrease the use of harmful pesticides that result in agricultural runoff. Urban and local farmers also eliminate the need for preservatives, as their products do not need to travel long distances.

Energy Efficiency

The current industrial agriculture system is accountable for high energy costs for the transportation of foodstuffs. The average conventional produce item travels 1,500 miles, using, if shipped by tractor-trailer, one gallon of fossil fuel per hundred pounds. The energy used to transport food is decreased when urban agriculture can provide cities with locally-grown food.

Economy of Scale

Using high-density urban farming, as for instance with vertical farms or stacked greenhouses, many environmental benefits can be achieved on a city-wide scale that would be impossible otherwise. These systems do not only provide food, but also produce potable water from waste water, and can recycle organic waste back to energy and nutrients. At the same time, they can reduce food-related transportation to a minimum while providing fresh food for large communities in almost any climate.

4

Rural Agricultural Development and Green Revolution

AGRICULTURE IN THE INDIAN STATES

About 70% population of the India is directly dependent on agriculture, also agriculture is the main component for most of the state economies in India. The Punjab region is suited for wheat-growing.

The south Indian states of Andhra Pradesh, Tamil Nadu and Karnataka are also major producers of rice. These states are mostly irrigated by the rivers Krishna, Godavari and Cauvery.

Haryana is self-sufficient in food production and the second largest contributor to India's central pool of food grains. Haryana contributed significantly to the Green Revolution in India in the 1970s.

The high-altitude states of Himachal Pradesh and Jammu and Kashmir are ideal for production of apples. Arunachal Pradesh also has a large number of fruit orchards.

Tea is the other produce of the high altitude regions of Assam, West Bengal (Darjeeling), Tripura, Tamil Nadu (Ooty), Himachal Pradesh and Kerala. Assam produces some of the finest and most expensive teas in the world (see Assam tea). However Assam tea is grown at elevations near sea level, giving it a malty sweetness and an earthy flavor, as opposed to the more floral aroma of highland (e.g., Darjeeling, Taiwanese)

In the north-eastern states like Arunachal Pradesh and Nagaland, shifting cultivation known locally as jhum, was practised by the tribal groups, but that has come to be less practised.

In mountainous states like Sikkim, farming is done on terraced slopes. The state has the highest production and largest cultivated area of cardamom in India. Karnataka is the largest producer of coffee in India and accounts for 59% of the country's coffee production.

It is grown mostly in the lower slopes of the Western Ghats in Kodagu district. Karnataka is also the largest producer of sandalwood based products like perfumes. Sandalwood comes from the dense forested areas of southern Karnataka.

The state is also the largest producer of raw silk and Mysore silk saris are famous. The other state manufacturing silk is Tamil Nadu. Rajasthan is among the largest producers of edible oils in India and the second largest producer of oilseeds.

The state is also the biggest wool-producing state in India. Kerala produces 96% of national output of pepper. It is also the largest producer of spices which include cardamom, vanilla, cinnamon, and nutmeg. Kerala is also the largest producer of natural rubber in India (91%). The other state producing rubber is Tripura.' 30 percent of Mizoram is covered with wild bamboo forests, and accounts for 40 percent of India's 80-million-ton annual bamboo crop. There has been an economic shift towards agriculture for some states like Bihar after the mineral rich areas were carved out to form the new state of Jharkhand.

Research

The Indian Agricultural Research Institute (IARI) established in 1905, was responsible for the research leading to the "Green revolution" of the 1970s. The Indian Council of Agricultural Research (ICAR) is the apex body in agriculture and related allied fields, including research and education. The Union Minister of Agriculture is the President of the ICAR. The Indian Agricultural Statistics Research Institute develops new techniques for the design of agricultural experiments, analyses data in agriculture, and specializes in statistical techniques for animal and plant breeding.

India is now trying to adopt organic farming, but as per the survey conducted in the year 2006-2007, the total production and the total requirement of organic manures and biofertilizers in India was estimated approximately to be 12.2 million tonnes and 15.7 million tonnes respectively. Hence, the research is now focussing towards minimizing the gap between the supply and the demand of organic manures and biofertilizers. Prof. M.S. Swaminathan is known as "Father of the Green Revolution" and heads the MS Swaminathan Research Foundation. He is known for his advocacy of environmentally sustainable agriculture and sustainable food security.

ALL INDIA PRODUCTION OF FOODGRAINS (JULY-JUNE)				
			(IN MILLION TONNES)	
GRAINS	**2001-2002***	**2002-2003***	**2003-2004@**	**2004-2005***
		(FINAL)		
RICE				
KHARIF	80.52	63.08	78.24	73.29
RABI	12.82	8.74	10.04	14.51
TOTAL	93.34	71.82	88.28	87.80
WHEAT				
RABI	72.77	65.76	72.11	73.03
COARSEGRAINS				
KHARIF	26.71	19.99	32.37	24.60
RABI	6.66	6.08	5.75	7.28
TOTAL	33.37	26.07	38.12	31.88
TOTAL CEREALS				
KHARIF	107.23	83.07	110.61	97.89
RABI	92.25	80.57	87.90	94.84
TOTAL	199.48	163.64	198.51	192.73
PULSES				
KHARIF	4.84	4.15	6.16	5.04
RABI	8.53	6.97	8.78	8.63
TOTAL	13.37	11.12	14.94	13.67
TOTAL FOODGRAINS				
KHARIF	112.07	87.22	116.77	102.93
RABI	100.78	87.55	96.68	103.45
TOTAL	**212.85**	**174.77**	**213.45**	**206.38**

(SOURCE : DIRECTORATE OF E & S)
@ 2nd Advance Estimates as on 5.8.04
(*) First Advance Estimates as on 19-01-2005

AGRICULTURE AND FOOD SECURITY

Pre-Independence India suffered repeated famines, drought and food shortages. But following the Green Revolution in the '60s, yields and foodstocks rose manifold. Now, 30 years later, Indian farmers have realised the follies of their tryst with intensive agriculture. Despite 70 per cent of the population being engaged in agriculture and allied activities, declining foodgrain production and access to food remain the two biggest problems confronting the country. Liberalisation has made things worse: commercial crops are eating into the fertile land tracts meant for essential foodgrains. And six years after the World Trade Organisation came into existence, the anticipated gains for India from the trade liberalisation process in agriculture are practically zero.

Agriculture Minister C Subramaniam sat huddled with his senior staff. There was tension in the air. Depending on imported foodgrains coming from 12,000 miles away, particularly when it came in driblets following US President Lyndon Johnson's directive to "teach India a lesson", was humiliating enough. What worried him was that food stocks had reached such a precarious low that they were only sufficient for another two weeks. Still worse, there was nothing in transit. Recalls Subramaniam, "As a last resort, I told my officials and experts to identify the nearest food-carrying ships on the ocean throughout the world. I said we would identify those carrying wheat to other countries and appeal to the US President to divert them to India if other countries could wait for another six to eight weeks."

This was in 1966-67, the critical year of drought when India imported 11 million tonnes of foodgrain. A year earlier, India had imported 10 million tonnes of foodgrain. Around this time, the Paddock brothers, often referred to as 'prophets of doom', concluded in their book, Famine 1975, that by the mid-'70s at least half of India would be led to a slaughter house. Considering that India had a long history of hunger and famine, and the subsequent failure of the government's effort to boost agricultural production in the post-Independence era, not many disagreed with the conclusive analysis drawn by William and Paul Paddock.

After all, for a country literally surviving on a 'ship-to-mouth' existence, there was little hope. Soon after Independence, India had begun to seek food aid and in fact emerged as the biggest food importer of this century. In 1951, India got ad hoc assistance of 2 million tonnes of foodgrain to tide over the crisis arising from severe drought in several parts of the country. This was the first major food security on a significant scale after Independence.

Before 1947, Indian history was replete with famine, drought and food shortages. Between 1770 and 1880, as many as 27 food scarcities and famines were recorded. At least 20 million lives were lost in India in about 20 famines that had struck since 1850.

Much of this loss was because of misplaced colonial policies which aimed to derive maximum economic gain at the cost of human suffering and misery. After the British had created a transport infrastructure in the first half of the 19th century, they began encouraging farmers to grow crops that could be exported.

The boom in export trade accompanied by rising prices made farmers shift to cash crops like cotton, indigo, poppy and sugarcane. The area under foodgrain subsequently shrank. In other words, efforts to improve agriculture in colonial India were directly linked to the needs of British industries.

In the first half of the 20th century, India's foreign trade showed the same colonial pattern with exports mainly comprising foodgrain and cotton, jute, oilseeds, opium and indigo. Against this, imports consisted mainly of consumer goods. Meanwhile, the cultivator's preference for commercial crops continued.

As is happening now, the trend during the pre-Independence years showed that the best lands available were used for the cultivation of commercial crops. The cultivation of foodgrains was relegated, to some extent, to inferior lands, the yield of which is much less even in normal years. The result was a steady decline in foodgrain output till the Second World War. Per capita food availability from domestic production too had declined. As B M Bhatia observes in The Famines of India: "India kept the precarious balance between increasing population and declining food production partly by turning from an exporter into a net importer of foodgrains but largely by the easy, though dangerous, expedient of pushing down the already low per capita cereal consumption to still lower levels."

The big gap between minimum requirement and supply continued throughout the post-War period. Till the 1960s, Indian agriculture was not sufficiently able to meet domestic requirements and the country had to rely on foodgrain imports.

SOWING THE SEEDS OF THE GREEN REVOLUTION

Realising the disastrous consequences of the growing gap between the rates of growth of population and food production, a vigorous 'grow more food' campaign was launched. The key goal of economic planning shifted to attaining self-sufficiency. The First Five-Year Plan, following food imports in 1951, gave maximum importance to the growth of agriculture. Among the measures initiated in the '50s to stimulate foodgrain production were land reforms, irrigation, fertiliser production, strengthening of research and the organisation of a national farm extension service. Consequently, increased harvests in the mid-1950s relieved the economy.

It was around this time that the government adopted the US Land Grant model of agricultural research. Agricultural universities, beginning with the Gobind Ballabh Pant University of Agriculture and Technology in Pantnagar in Uttar Pradesh, were gradually set up in all the states. The Land Grant model entailed linking research with extension, and taking research from the laboratories to the farms through an extension network linked with the university. In 1961, the government launched the Intensive Agriculture District Programme (IADP) to enhance productivity in the irrigated areas. Although the aim was to introduce improved seed along with a package of agronomic practices to take advantage of the available irrigation, the programme did not come up to expectations for want of high-yielding varieties of wheat.

The tall Indian wheat varieties did not provide an economic response to the application of chemical fertilisers and irrigation water, which agricultural scientists were desperately looking for. These had a tendency to fall down or lodge when fertilisers were applied, as a result of which the yields stagnated at less than one tonne per hectare. Even though breeding for improved non-lodging wheat varieties was initiated in the '50s, it did not meet with much success.

The missing link was provided by Dr Norman Borlaug, who was then working with CIMMYT in Mexico. A few dwarf wheat strains of spring wheat that he sent to the Indian Agricultural Research Institute in New Delhi, looked impressive. In the next two to three years, a wide range of dwarf material was tested under the All India Wheat Research Project. In August 1964, the then Professor and Head of Department of Plant Breeding and Genetics at the IARI, Dr M S Swaminathan, who later emerged as the architect of India's Green Revolution, proposed the launching of a National Demonstration Programme in farmers' fields. This would verify the results obtained in research plots and introduce farmers to new opportunities opened up by semi-dwarf varieties for considerably improving the productivity.

Says Swaminathan, "When small farmers, who with the help of scientists organised the National Demonstration Programme, harvested over 5 tonnes of wheat per hectare, its impact on the minds of other farmers was electric. The clamour for seeds began and the area under high-yielding varieties rose from 4 hectares in 1963-64 to over 4 million hectares in 1971-72. A small government programme became a mass movement." Subsequently, the government imported 18,000 tonnes of semi-dwarf wheat from Mexico in 1966 to be distributed in time for immediate sowing by farmers in the northwestern regions.

The seeds of the Green Revolution were, therefore, truly sown in the mid-'60s. What followed next is already part of contemporary history. Wheat production rose to 17 million tonnes in 1968, an increase of 5 million tonnes over the highest of 12 million tonnes harvested in 1964. Indira Gandhi officially recorded the impressive strides in agriculture by releasing a special stamp entitled 'Wheat Revolution' in July 1968. The success of wheat was later replicated in rice. Productivity increases were also recorded in cotton, sugarcane, millets and oilseeds.

MEASURES TO AVOID FAMINE

The famine-avoidance strategy that India launched had all the necessary ingredients for a sustained effort. In 1965, the Agricultural Prices Commission was constituted, with the basic objective of assuring fair prices for farm produce. The Commission, an autonomous body, works out the cost of cultivation for 22 agricultural commodities, and suggests a price to the government that includes profit that should serve as an incentive to grow more. The administered price of wheat, for instance, was raised by 71 per cent between 1972 and 1981. Similarly, the price of rice was jacked up by 81 per cent during the same period. Later, to sustain the tempo of the rice revolution, the relative price structure was moved in its favour. In the early phase of the Green Revolution, the price policy favoured wheat, but after the mid-'70s it tilted in favour of rice. For oilseeds and pulses, the reverse was put into action. Relatively high support prices were announced in the early-'90s to make their cultivation profitable.

Perhaps the most significant aspect of the price support mechanism has been the insulation of farmers against a decline in prices. Wheat production was very high in 1971-72, as a consequence of which prices should have declined by at least 25 per cent. Since the government purchased all the wheat that was offered, farmers were actually paid 7 per cent more. Again, in 1982-83, when wheat production was 14.3 per cent higher than the previous year, the prices should have actually declined by 32.5 per cent. With the government purchasing all the wheat that flowed into the markets, however, the farmers realised 18 per cent more.

The biggest achievement of the Indian food policy, dependent more and more on buffer stocking and operational stock holding, has been the avoidance of famine-like conditions. It was with the basic objective of curbing consumption and ensuring an equitable distribution of available food supplies, especially in the deficit areas and among the poorer strata of society, that the Public Distribution System (PDS) was introduced. PDS supplies were also used for the 'food for work' programme as well as other anti-poverty programmes. At present, PDS covers more than 80 million families and the total foodgrains distributed through a network of 40,000 fair price shops accounts for over one-third of the total trade in foodgrain.

AIMING AT SELF-SUFFICIENCY

The Green Revolution was certainly not a stroke of luck. The resilience was the result of progressive farm policies, determination, sustained efforts of scientists and the vigour of the Indian farmers who readily adopted the improved technology. This is what the Rockefeller Foundation had to say about the magnificent strides taken: " ... The speed with which India's farmers and scientists suddenly gave their country an approach to an abundant food supply, has never been duplicated on an equal scale anywhere in the world, including the scientifically sophisticated United States."

Since the early-'70s, the country has avoided famine even during adverse weather conditions, through a carefully designed food security system involving the maintenance of both substantial grain reserves and an extensive PDS. The devastating drought of 1987, rated as the most severe of this century, affected nearly 155 million hectares. In spite of this, no starvation and hunger deaths were reported. This was essentially the result of excellent food management based on foodgrain reserves built largely from homegrown wheat and rice.

From the late-'60s, the rate of growth in food production generally exceeded the rate of growth of population. Despite the continuous growth in population, the per capita availability of foodgrain increased from 157 kg per year in 1955 to 177 kg per year in 1995, an increase of 15.7 per cent. Food production has risen roughly four times since the days of Independence, from 50 million tonnes in 1950-51 to 198 million tonnes in 1996-97. At the same time, the population has risen from 300 million to 960 million. Milk production too has risen to 68 million tonnes, the highest in the world. Food self-sufficiency was achieved in the late-'70s with the maintenance of a sizeable food buffer. Lately, India has begun to export foodgrain. About 5 million tonnes of rice and wheat were exported during the last few years. Skimmed milk powder amounting to about 5,000 tonnes has also been exported.

The increase in foodgrains was, however, disproportionate. The highest rate of growth was achieved in Punjab and Haryana, where, at a compound growth rate of 4.63 per cent, foodgrain production jumped from 7.23 million tonnes in 1964-65 to reach an all-time high of 30.33 million tonnes in 1995-96. Uttar Pradesh recorded a compound growth rate of 3.01 per cent. Production in Maharashtra, Madhya Pradesh, Bihar, Orissa and the northeastern states continues to stagger. Tamil Nadu and Andhra Pradesh, on the other hand, have recorded significant increases in rice yields, with production rising to 9.17 million tonnes and 11.58 million tonnes, respectively.

The Green Revolution was aimed at the better-endowed regions. For millions of farmers languishing in the drylands, constituting more than 70 per cent of the cultivable lands, it continues to be a futile struggle. Despite emphasis on dryland farming during the past several decades, the scenario still remains grim. The undulating topography and the irregular rainfall pattern have combined to aggravate the situation. That the drylands produce about 42 per cent of the country's food shows that the future of farming lies in these areas. Nearly 83 per cent of sorghum, 81 per cent of pulses and 90 per cent of oilseeds grown in the country come from these areas. The poor yields and the fluctuations in production are indications of the scant attention drylands have received from policymakers and planners.

The problem of increasing productivity on drylands has serious socio-economic implications. With every passing year, the gap between the farmer's yields in irrigated areas and in the dry farming region is widening. One year of drought is enough to push a farmer into a deep well of poverty for another two to three years. Drought is a recurring phenomenon in arid and semi-arid areas. Fifty years after Independence, life for millions of people somehow surviving in the drylands continues to be worse than before.

DECLINE IN PRODUCTION

Thirty years after the dawn of the Green Revolution, Indian farmers are realising that their love affair with intensive agriculture is on the decline. Despite a bountiful monsoon for the tenth year in a row, harvests are not as plentiful as expected. However, with no let-up in the ever-increasing rise in population, Indian agriculture is once again at the crossroads.

Alarm bells have been ringing for quite some time. For nearly a decade, agricultural production had stagnated. The spectacular yield growth recorded in the post-Green Revolution years in Punjab and Haryana has receded into history. Of the multiple problems confronting agriculture, rapid fragmentation of land holdings is keeping pace with increasing population. In 1976-77, the average size of the holdings was estimated at 2 hectares, while in 1980-81, it came down to 1.8 hectares. Today, it stands at a mere 0.2 hectares. The total number of land holdings in 1981 were around 89 million; today these have crossed 100 million. By the turn of the century, the average land holding will come down to 0.11 hectares. It is quite obvious that with such small land holdings, Indian agriculture cannot adopt high-tech farm practices.

Basking in the afterglow of the Green Revolution, farming and agriculture have ceased to attract serious attention. In 1995-96, foodgrain production slumped to minus 3.60 per cent. In 1997-98, it further declined to minus 3.70 per cent, the worst-ever since the heady days of the Green Revolution. Foodgrain production in the frontline agricultural states of Punjab, Haryana and western Uttar Pradesh, comprising the country's food bowl, has decelerated. The miracle that began with wheat was replicated in rice. This is where the cost-benefit ratio has turned negative. The downward spiral in food production continues through the southern regions of the country. Tamil Nadu, another Green Revolution area, is under tremendous strain from intensive cultivation. In Karnataka, the negative trend in yield levels of all food crops, barring cotton and sugarcane, are all too apparent. Farming in Karnataka can be clearly separated in two distinct classes, the 'creamy layer' of corporate agriculture occupying the fertile and irrigated areas, and the remaining low productive tracts at the mercy of the rains, constituting the tiny and small land holdings. Kerala too, picturesque and verdant with tropical forests, is passing through its worst-ever crisis on the foodgrain front.

As if this is not enough, the number of landless in rural areas too has been multiplying over the past few decades, at an estimated rate of two million every year. The rapid expansion in the number of land holdings, with less than a hectare under plough, has meant an additional 6,00,000 farmers every five years or so. The negative terms of trade for agriculture and the declining public sector investments in farming are indices of the sluggishness in this sector.

The Economic Survey for 1996-97 has further sounded a warning. The compound growth rate in foodgrain production at 1.7 per cent between 1991 and 1996, is lower than the annual population growth of 1.9 per cent. The survey adds that even a marginal fall of three to four per cent in foodgrain output can cause the prices of essential commodities to escalate sharply, necessitating government intervention by way of resorting to imports. This happened in 1993 and again in 1996.

The philosophy of agricultural planning is changing. Gone are the days when the nation's emphasis was solely on attaining self-sufficiency in foodgrain production. Gone also are the days when farmers were newly independent India's heroes, revered for their role in keeping hunger and starvation at bay. Today, at a time when food production struggles to keep pace with the burgeoning population growth, farmers are being asked to diversify, produce crops that are suitable for export and to compete in the international market. With the promise of cheap food available off the shelf in the global market, the focus has shifted from agriculture to industry, trade and commerce, from the small and marginal farmers to the agri-processing companies, which alone can bring in investments and add value to produce.

The restructuring of the Indian economy too has put the emphasis on agri-exports. Cultivation of staple food is being replaced by cash crops, tomatoes in place of wheat, durum wheat (for bakery purposes) replacing wheat as a staple diet in Punjab and Haryana, flowers in place of rice, and so on. In the coastal areas, private enterprise is taking away the fish catch, depriving the local communities of a livelihood and their only source of nutrition. In Kerala, vast tracts of forests and paddy fields have been converted into rubber, coffee and coconut plantations. Every year, about 25,000 hectares of good paddy land are being diverted for non-paddy purposes. The structural transformation is not only peculiar to Kerala. It is happening in almost all the states. Commercial crops are eating into the fertile land tracts meant for growing essential foodgrains, thereby making the states rely more heavily on the PDS. The diversion of good agricultural land, which in any case is limited, to commercial farming and even industries, is further restricting the ability to grow enough foodgrain.

IN THE ERA OF WTO

Kerala has already been at the receiving end. Over the past few years, ever since economic liberalisation became the development mantra, the state has been flooded with cheap and highly subsidised agricultural imports, throwing its agrarian economy out of gear. Whether it is the import of palm oil or rubber or coffee, almost every aspect of the state's socio-economy has been negatively impacted.

Coconut prices have crashed, down from Rs 10 to Rs 2. Rubber has plummeted from Rs 60 to Rs 16 and coffee from Rs 58 in 1999 to Rs 30 per kg in 2000. Even spices have not been spared, with pepper prices falling from Rs. 2,600 to Rs. 1,300 per quintal in the consecutive period. Kerala is not alone.

The fallout from the emerging global trade paradigm is being felt all over the country, though not in the same magnitude. Before we go any further however, it is important to understand how the market rules play, and then try to understand the implications for Indian agriculture.

It is now official. Six years after the World Trade Organisation (WTO) came into existence on January 1, 1995, the anticipated gains for India from the trade liberalisation process in agriculture are practically zero. The ministry of agriculture as well as the ministry of commerce have admitted that the hopes from an international regime that talked of establishing a fair and market-oriented agricultural trading system have been belied.

It is time to draw a balance sheet of the gains and losses that accrued from the implementation of the WTO's Agreement on Agriculture, which had incorporated three broad areas of commitments from member states, namely in market access, domestic support and export subsidies. The underlying objective of this was to correct and prevent restrictions and distortions in world agricultural markets. Six years later, it has been established that these measures have only protected the farmers and the farming systems of the developed countries. On the other hand, the trading regime has ensured that developing countries take time-bound initiatives to open up their domestic markets for cheap and highly subsidised imports of agricultural commodities.

EXPORT SUBSIDIES URGED

The agriculture ministry acknowledges that despite the rules being defined, the expected gains have eluded the developing countries. It was expected that with the removal of trade distorting measures, agricultural exports from the developing countries would increase. This did not happen. In fact, India has seen a three-fold increase in the imports of agricultural commodities and products, from about Rs 5,000 crore in 1995 to over Rs 15,000 crore in 1999-2000. In edible oils alone, the import bill has soared to Rs. 9,000 crore. The so-called fair trading system has also not helped efficient producers in realising a higher price for their products. On the contrary, prices of most agricultural commodities are declining in the world markets.

In conclusion, if declining foodgrain production and access to food remain the two biggest problems confronting the country, there must be something terribly wrong with the way we look at agriculture. With more than 70 per cent of the population still engaged in agriculture and allied activities and an equal percentage of farmers tilling an average of 0.2 hectares of land and somehow surviving against all odds, the time has come to set the balance right. Whether we accept it or not, India is gradually moving back to the pre-Green Revolution days of a 'ship-to-mouth' existence, when food was largely imported to feed the hungry. It was the political maturity of the then leadership that led to self-sufficiency on the food front. Few will still question what Jawaharlal Nehru once said: "Everything else can wait, but not agriculture."

INDIA: AGRICULTURE AND RURAL DEVELOPMENT

From a nation dependent on food imports to feed its population, India today is not only self-sufficient in grain production but also has a substantial reserve. The progress made by agriculture in the last four decades has been one of the biggest success stories of free India. Agriculture and allied activities constitute the single largest contributor to the Gross Domestic Product, almost 33 percent of it. Agriculture is the means of livelihood of about two-thirds of the workforce in the country.

This increase in agricultural production has been brought about by bringing additional area under cultivation, extension of irrigation facilities, the use of improved high-yielding variety of seeds, better techniques evolved through agricultural research, water management, and plant protection through judicious use of fertilisers, pesticides and cropping practices.

Crops

The 1970s saw a multi-fold increase in wheat production that heralded the Green Revolution. In the next decade rice production rose significantly; in 1995-96, rice production was 79.6 million tonnes. The total grain production crossed 211 million tonnes in 2001-02, a big leap from 51 million tonnes in 1950-51. To carry improved technologies to farmers, a National Pulse Development Programme, covering 13 states, was launched in 1986. Efforts to boost pulse production were augmented further by the Special Food Production Programme. In 2001-02, pulse production was 13.52 million tonnes. With some states offering more than the statutory minimum price, sugar cane production also received a boost, and in 2001-02 a record 292.2 million tonnes was registered.

Irrigation

As efforts continued to increase the irrigation potential in the country, the last 40 years saw the gross irrigated area reach 85 million hectares. Flood forecasting has become an important activity over the years. Over 500 hydrological stations collect and transmit data through 400 wireless stations for issuing forecasts for 157 sites. About 5,000 forecasts are issued in a year with 94 percent accuracy. The country also receives international support, with the World Bank as a primary source, for developing its water resources. International cooperation is also envisaged in setting up a National Centre for Information on Water and Power. As there is a broad seismic belt in the country, particularly along the Himalayan, and the Kutch region and parts of Maharashtra, a scheme is being evolved to collect all data on seismic activity at various dam sites.

Fertilisers

The fertiliser industry in India has grown tremendously in the last 30 years. The Government is keen to see that fertiliser reaches the farmers in the remote and hilly areas. It has been decided to decontrol the prices, distribution and movement of phosphatic and potassic fertilisers. Steps have been taken to ensure an increase in the supply of non-chemical fertilisers at reasonable prices. There are 66 fertiliser quality control laboratories in the country. Since biofertilisers are regarded as an effective, cheap and renewable supplement to chemical fertilisers, the Government is implementing a National Project on Development and Use of Bio-fertilisers. Under this scheme, one national and six regional centres for organising training, demonstrating programmes and quality testing of bio-fertilisers has been taken up.

It was a challenging decision of the Government to take Bombay High gas through a 1,700 km pipeline to feed fertiliser plants located in the consumption centres of North India. However, the major policy which has ensured the growth of the fertiliser industry is the thrust on accelerating fertiliser consumption by fixing, on the one hand, low and uniform price for fertilisers, and on the other hand providing the manufacturers adequate compensation through the retention price and subsidy scheme. As expected, fertiliser nutrient demand has gone up from 0.29 million tonnes in 1960-61 to 16.7 million tonnes at the end of 2000-01, compared to 12.15 million tonnes during 1992-93.

Fisheries

Fish production achieved an all-time high of 5.6 million tonnes at the end of 2001-02. Programmes that have helped boost production include the National Programme of Developing Fish Seeds, Fish Farmers' Development Agencies and Brackish Water Fish Farmers' Development Agencies. The Central Institute of Fisheries Nautical and Engineering Training trains the necessary manpower. To diversify fishing methods and introduce processed fish products on a semi-commercial scale, an Integrated Fisheries Project has been launched. A National Fisheries Advisory Board has also been established.

AGRICULTURAL RESEARCH

The apex body for education, research, training and transfer of technology in the field of agriculture is the Indian Council of Agricultural Research (ICAR), established in 1929. India's transformation from a food deficit to a food surplus country is largely due to ICAR's smooth and rapid transfer of farm technology from the laboratory to the land.

ICAR discharges its responsibilities through 43 research institutes, four national research bureaux, 20 national research centres, nine project directorates,

70 all-India coordinated research projects, and 109 Krishi Vigyan Kendras (farm science centres). Besides, the programme of Agricultural Education is coordinated by ICAR with the curricula and other normative guidance given to the 26 agricultural universities and four national research institutes.

OILSEEDS PRODUCTION

A Technology Mission on Oilseeds was launched in 1986 to increase production of oilseeds in the country and attain self-sufficiency. Pulses were brought under the Technology Mission in 1990. After the setting up of the Technology Mission, there has been consistent improvement in the production of oilseeds. The oilseed production, which was 108.3 lakh tonnes in 1985-86, has increased to 247.30 lakh tonnes in 1998-99, the highest production so far. However, due to unfavourable weather conditions prevailing in the major oilseeds-growing states, the oilseed production during 1999-2000 and 2000-01 suffered some decline.

The increase in production was largely contributed by soyabean, rapeseed and mustard. Production of pulses has seen many ups and downs, which is expected to be checked under the mission. The country grows mainly nine oilseeds, with groundnut, rapeseed and mustard accounting for 62 percent of total production. Lately, soyabean and sunflower have shown major growth potential.

DRINKING WATER

A Technology Mission on Drinking Water and Related Water Management has been constituted to cover the residual problem villages and provide potable water at 40 litres per capita per day, and 70 litres per capita per day in desert areas inclusive of 30 litres for cattle. The mission is tackling the problem through 55 mini-missions in project districts and countrywide problem-oriented sub-missions. A Village Level Operation and Maintenance (VLOM) pump called India Mark-II has been developed and is being exported to 40 countries.

AGRICULTURE IN INDIA

Agriculture in India is one of the most prominent sectors in its economy. Agriculture and allied sectors like forestry, logging and fishing accounted for 18.6% of the GDP in 2005 and employed 60% of the country's population. It accounts for 8.56% of India's exports. About 43% of India's geographical area is used for agricultural activity. Despite a steady decline of its share in the

GDP, agriculture is still the largest economic sector and plays a significant role in the overall socio-economic development of India. The monsoons play a critical role in the Indian sub-continent's agriculture in determining whether the harvest will be bountiful, average, or poor in any given year. The entire rainfall in the sub-continent is concentrated in the few monsoon months. Agriculture in India is constitutionally the responsibility of the states rather than the central government. The central government's role is in formulating policy and providing financial resources for agriculture. The government administers prices of essential commodities to protect farmer's interests. It also administers other commodities which are produced by government-controlled companies, like petroleum, coal, nitrogenous fertilizers, etc. Other than these, most agricultural commodity markets operate under the normal forces of demand and supply.

HISTORY OF INDIAN AGRICULTURE

Evidence of the presence of wheat and some legumes in the 6th millennium BC have been found in the Indus Valley. Oranges were cultivated in the same millennium. The crops grown in the valley around 4000 BC were typically wheat, peas, sesame seed, barley, dates and mangoes. By 3500 BC cotton growing and cotton textiles were quite advanced in the valley.

By 3000 BC farming of rice had started. Other monsoon crops of importance of the time was cane sugar. By 2500 BC, rice was an important component of the staple diet in Mohenjodaro near the Arabian Sea. The Indus Plain had rich alluvial deposits which came down the Indus River in annual floods. This helped sustain farming that formed basis of the Indus Valley Civilization at Harappa. The people built dams and drainage systems for the crops. By 2000 BC tea, bananas and apples were being cultivated in India. There was coconut trade with East Africa in 200 AD. By 500 AD, egg plants were being cultivated.

Agricultural Movements

Green Revolution

After independence, considering India's growing population, the government took steps to increase the food production. Yields per unit area of all crops have grown since 1950. The 1970s saw a huge increase in India's wheat production. This is known as the Indian Green Revolution. Reasons for the growth are the special emphasis placed on agriculture and steady improvements in irrigation, technology, application of modern agricultural practices and provision of agricultural credit and subsidies

Operation Flood

Operation Flood was the name of a rural development programme started by the National Dairy Development Board (NDDB) in 1970 with the objective of creating a nationwide milk grid. This movement followed the Indian green revolution and helped in alleviating poverty and famine levels from dangerous proportions in India during the era. It resulted in India becoming the largest producer of milk and milk products, so it is also called the White Revolution of India.

Production

India is ranked second in the world in terms of agricultural output. India is the largest producer in the world of milk, cashew nuts, coconuts, tea, ginger, turmeric and black pepper. It also has the world's largest cattle population (193 million). India ranks second worldwide in farm output. It is the second largest producer of wheat, rice, sugar, groundnut and inland fish. It is the third largest producer of tobacco. India accounts for 10 per cent of the world fruit production with first rank in the production of banana and sapota. Given below is a chart of trend of output of cereals and major foodgrains as published by the Department of Food and Public Distribution with figures in tonnes.

Year	Cereals	Rice	Wheat	Coarsegrains	Pulses
2001–02	199,480,000	93,340,000	72,770,000	33,370,000	13,370,000
2004–05	192,730,000	87,800,000	73,030,000	31,880,000	13,670,000

Despite high growth, international comparisons reveal that the average yield in India is generally 30% to 50% of the highest average yield in the world.

Productivity

The following table shows the harvest per acre of farmland in 2007 with figures in tonnes.

Commodity	Output per acre
National Average	0.29
Sugarcane	26.15
Potato	6.32
Wheat	0.84
Rice	0.70
Corn	0.64
Groundnut	0.37
Soyabean	0.32

The low productivity in India is a result of the following factors:

- Illiteracy, reforms and inadequate or inefficient finance and marketing services for farm produce.
- The average size of land holdings is very small (less than 20,000 m^2) and are subject to fragmentation, due to land ceiling acts and in some cases, family disputes. Such small holdings are often over-manned, resulting in disguised unemployment and low productivity of labour.
- Adoption of modern agricultural practices and use of technology is inadequate, hampered by ignorance of such practices, high costs and impracticality in the case of small land holdings.
- Irrigation facilities are inadequate, as revealed by the fact that only 53.6% of the land was irrigated in 2000-01 which result in farmers still being dependent on rainfall, specifically the Monsoon season. A good monsoon results in a robust growth for the economy as a whole, while a poor monsoon leads to a sluggish growth. Farm credit is regulated by NABARD, which is the statutory apex agent for rural development in the subcontinent.

In the last few decades several farmers have committed suicide especially in the states of Andhra Pradesh, Maharashtra, Karnataka, and Kerala. Combating this has become a major challenge for these governments. Some of the causes for the deaths include indebtedness of small and marginal farmers and repeated crop failures.

Waste

As per the report of the 11th Planning Commission, preventable post-harvest losses of foodgrains in India were estimated at about 20 million tonnes a year, which is nearly 10.5 percent of the total production due to inadequate storage, security and distribution facilities.

5

Agricultural Performance in Sustainability Development

AGRICULTURAL DEVELOPMENT IN SUSTAINABILITY

Agricultural development has three distinct but related dimensions: the physical-technical, the economic-financial, and the institution-human. The physical-technical dimension addresses land utilization, agricultural technologies, research and extension, agricultural inputs, farm-to-market access, productivity and production maximization, and so on. Agricultural development from the economic-financial perspective is concerned with costs, factors of production, terms if trade, pricing policies, subsidies, incentives, credit, return on investment, market mechanisms, and the like.

The institutional-human dimension looks at knowledge and skills, organization and management, training, implementation capacity, social relations, politics, communication, motivation, participation, local government, public-private sector linkages, culture and values, historical experience, and so on. Agricultural development, as with much of the development field, has tended to be the province of the technical and economic specialists. Their language and concerns have dominated the analysis of rural and agricultural sector issues, the specification of agricultural development objectives, and the design of programmes and projects to achieve those objectives. The institutional-human dimension has frequently been either completely ignored or treated as a source of problems or constraints to achieving technical and economic targets, a residual category for anything not defined as technical or economic.

For example, a World Bank staff member notes that, "between 1976 and 1980 the Bank invested U.S. $920 million in agricultural projects without funds in any one of them being allocated to the staff training needed for the new activities on which those projects depended".

Recent work by Ruttan represents one effort to build theoretical integration of the three dimensions. On the applied front, the USAID-supported farming systems approach to agricultural development is one example of an explicit attempt to integrate the three dimensions operationally.

Recognition of the importance of institutional and human factors in socio-economic development in general has increased substantially over the past 15 years or so, and is now a high priority area of attention by donor agencies and developing country governments alike. Being a social technology, institutional development does suffer in the eyes of crop scientists, agricultural economists, and macroeconomic analysts from a lack of specificity, predictability, and hard-and fast rules.

Despite its "softness," institutional development is a frequently espoused need of developing country officials and an aim often expressed by donor agencies. U.S. foreign assistance policy, for example, aims officially "to build and maintain the social and economic institutions necessary to achieving self-sustaining growth".

This goal of institutional development is nowhere more important than in the rural and agricultural sectors, which continue to be the source of most employment in developing countries. Except for emergency relief, where the delivery of food for consumption is unavoidable, donor policies stress that assistance for the countryside should be an investment, that it generate a flow of benefits that endures after the external funds run out.

The cumulative lessons of experience show that to have lasting impact, donor efforts to improve rural areas in the developing countries, and agriculture in particular, need to work with durable indigenous institutions both at the central and local levels. It is not surprising that, in the 1990s, institutional sustainability is emerging as one of the major problem areas for sectoral management, as well as for public administration of core government functions in developing nations.

This chapter introduces the topic of institutional sustainability, focusing in particular on the agricultural and rural sectors, though its observations apply equally to other parts of the economy. To set the stage for the rest of the book, we start by briefly exploring the meanings of the key terms "sustainability" and "institution." Then we review the latest quantitative studies of sustainability, putting them in a comparative perspective.

The broad topic of development "sustainability" is obviously not a new issue. The folklore of foreign aid is replete with tales of equipment that never worked in the field and was left to rust, of well-intended recommendations that local people disregarded, of organizations that succumbed to apathy — if they ever functioned in the first place.

Critics of development assistance have always complained about this sort of waste. It has often been easier to be clear about the failed and was not sustained than to search out what worked, what lasted, and why. In the mid- to late 1980s, however, sustainability has surfaced as a distinct and pressing concern of policy-makers in donor agencies and in the developing world.

As the last decade of the twentieth century gets underway, the expansion of cropped area and the adoption of modern farming techniques are putting new, perhaps irreversible, stress on the natural environment in developing countries, while soft international markets for agricultural commodities are forcing reappraisal of the possibilities for the Third World to increase its farm exports. Conventional development strategies seem more and more to have run their course, yet fresh approaches have been difficult to find.

"Aid weariness" among the donor countries and high levels of indebtedness among the recipients have constrained the search for creative solutions. Finding ways to yes both external and indigenous development resources more efficiently to achieve broader and more lasting impact has become, accordingly, more difficult and more imperative.

Each of the academic disciplines cuts into sustainability problems from a different angle. Three points of view stand out, each reflective of the three dimensions of agricultural development previously noted. Agricultural and natural scientists tend to frame the problem of sustainability in terms of the long-term impact of current farming practices on the global resource base.

The chemicals and intensive cultivation associated with green revolution technology can degrade the environment and waste energy resources; scientific breeding of seed may reduce the native genetic diversity of crops, making food supplies more vulnerable to plant disease and pests. Because of such problems, agriculturalists and naturalistst are increasingly concerned with identifying and promoting methods of cultivation that can protect and enhance nature's assets, while continuing to produce enough food to satisfy burgeoning populations.

Thus the blue-ribbon Bruntland Commission has defined sustainable development as "development that meets the needs of future generations without compromising the ability of future generations to meet their own needs". Countries such as Nepal and Haiti are living (or dying) examples of unsustainable development, and represent cautionary stories for those who would ignore the ecological costs of detrimental policies and practices.

A downward spiral of events can result, with deforestation and erosion leading to food insecurity, extremes of poverty, class warfare, and ultimately to societal breakdown. Sub-Saharan African nations also appear to be in danger of proceeding down this road.

Economists, by contrast, are somewhat less preoccupied with the ecological dimensions of sustainability per se, since the mining of natural resources can be justified from an economic point of view. Their concern is not conservation by itself, but the appropriate tradeoffs between economic growth and environmental preservation.

Accordingly, economists tend to be troubled by the manmade market distortions that curtail the growth of production and employment in the Third World. The most widespread examples of these distortions are the ill-considered of politically convenient price regimens that penalize the rural sector in many developing countries. There are extended economic costs to such policies.

As the World Bank notes: "Discrimination against agriculture on a sustained basis not only reallocates resources within agriculture but also draws them out of it. As labor and capital move out and technical progress slow, the longterm losses can be large."

Management experts tend to make the development project or the organization their unit of analysis, and to worry about sustainability in terms of how to meet recurrent costs or permanently improve capacity for implementation. Development project management has emphasized achieving planned targets within a specified time frame and budget, thus responding more to donor agency needs for accountability than to recipient country capacity-building requirements.

Donor-supported projects, with their autonomous implementation units, frequently experience a collapsing "balloon effect" when external funds and advisers are withdrawn. Because they are often organizationally "quarantined" and protected, with special budget and staff, such projects may leave little lasting imprint on regular government agencies, and in many cases may actually weaken them by drawing off the best national talent.

The management perspective on sustainability, however, has evolved away from a narrow project focus, expanding to concentrate more on programmes integrated into developing country agencies and on the question of how to maintain an ongoing flow of goods and services to intended beneficiaries. From the administrative point of view, sustainability means building organizational capacity to perform over an extended time period (longer than the life of-project horizon) and assuring the continuation of useful and valued outcomes for client groups.

Specialists concerned with nonagricultural sectors cut into sustainability issues in their own way. Health experts, for example, are usually more concerned with the clientele than the implementing agency; they see the problem in terms of promoting affordable preventative health measures, a opposed to curative care alone, measures that enable poor people to enhance their physical well-being over the long run.

But the fundamental problems of sustainability in health and other fields are closely parallel to those facing farmers and rural administrators. The thread that connects the agro-environmental, economic, and management perspective on sustainability is effective social institutions, which much be in place to deal with all the current challenges to development.

Overcoming the environmental damage of modern farming, for instance, is likely to require (among other things) that scientific organizations attack novel research problems, that private enterprise produce greater quantities of more sophisticated farm inputs, that farm advisory services disperse new knowledge, that local authorities mobilize farmers to change collective and individual behaviour, and that farmers' organizations collect and market unfamiliar crops.

Even something as seemingly straightforward as price as price reform requires considerable institutional capacity to analyze and monitor the impact of new prices, let alone to manage any political backlash. The institutional dimension, for example, has emerged as critical to the success of the structural adjustment programmes and sector adjustment loans provided by the International Monetary Fund and the World Bank.

In virtually all problems of "development sustainability" the institutional dimension plays a key role. Unfortunately, many Third World countries lack the depth of institutions to carry through the complex and exacting duties likely to be needed to tackle the development tasks of the 1990s.

Rule-oriented Institution

What are institutions? The concept is subtle and thus subject to confusion. According to Huntington definition, they are stable, valued, recurring patterns of behaviour. Institutions thus include rules or procedures that shaped how people act, and roles or organizations that have attained special status or legitimacy. Their importance is now being rediscovered by the social sciences, including the development subfield.

An example of a rule-oriented institution is a system of land tenure; whereas a role oriented institution could be the legal authority established to adjudicate disputes arising out of the land-tenure system. Both rules and roles can be institutionalized, the former as codes of law or custom, the latter as concrete organizations.

It is useful to keep in mind the distinction between these two types of institutions. Development assistance for agriculture and rural development is sometimes aimed at altering a village society's fundamental rules, for instance, by promoting tenure reform and land redistribution. When practitioners seek to build better institutions, however, the role-oriented, organizational definition is usually what they have in mind.

The USAID-sponsored institution-building literature of the 1960s, for instance, converged around organizations makes this matter of official policy, arguing that "because institutions become tangible only through the policies and actions of particular organizations, much of [our] institutional development effort will be focused on improving the policies and procedures of key organizations."

Most other donors and developing countries adhere to a similar definition, often narrowing it to mean formal collectives in the public sector, such as ministries, parastatals, agencies, or commissions. This book follows these conventions part way, by focusing on institutions in the organizational sense; however, our framework of analysis allows, indeed requires us, to simultaneously consider the effect on organizations of established codes of behaviour.

The development community's focus on role-oriented, organized activities unfortunately introduces another point of confusion, since "institutions" in this sense can be used loosely to refer to any formal or semi-formal collective entity. But not all organizations are institutions, any more than all institutions are organizations. As Perrow points out: "Some organizations are merely... rational tools in which there is little personal investment and which can be cast aside without regret.

Others become institutionalized. They take on a distinctive character; they become prized in and of themselves, not merely for the goods or services they grind out. People build their lives around them, identify with them, become dependent on them." Many rural organizations fail this test. When one refers to them as institutions it is often to speak hopefully about what they might become, and not realistically about what they, in fact, are.

A prime example of this failure is the experience of many developing countries with agricultural cooperatives, where the gap between what the coops were intended to be, and what they actually are, has been wide.

Our contributors concentrate on formally-constituted collectivities, but without prejudging whether they deserve the appellation "institution" in a strict sense. Further, the fact that the case studies tend to neglect informal modes of social organization, should not be interpreted as indicating that such institutions are less important in development. Nor should the focus of the case studies on discrete institutions be taken to mean that one may ignore the contextual factors within which organizations functions. To understand sustainability, it is important to keep this larger environment clearly in view.

AGRICULTURAL GROWTH AND PERFORMANCE

While the above analysis has provided a general view of the impact of economic reforms, this chaper examines agricultural growth and performance in the states of Bihar, Karnataka, Tamil Nadu and Punjab with their attendant policy implications.

The yields for various crops in these states differ greatly. While Tamil Nadu had the highest yield in rice, oil seeds and sugarcane, Punjab enjoyed the highest yields in wheat, coarse cereals, pulses and food grains. Karnataka on the other hand is seen to do well in cotton and Bihar performed quite well in pulses and coarse cereals. Further analysis and findings by Kalirajan and others (2001) show that Punjab had made remarkable achievements on the agricultural front while Bihar had remained stagnant in the last two decades, with Karnataka and Tamil Nadu showing moderate achievement. Clearly, differences in physical endowments, climatic conditions and institutional characteristics are some of the reasons for the varying productivity performance. Thus, having across the board economic reforms is likely to work less effectively than state-specific policy measures that enable each state's agricultural yields to reach their full potential. The comparative advantage of each state's agricultural production should be determined and with inter-state restrictions removed, total agricultural output would see a very significant increase.

For example, Karnataka with less favourable soil and water resources should be given incentives to concentrate on agro-processed products and corporate agriculture in horticulture, floriculture and animal husbandry, or to undertake watershed development to help with dry land agriculture. Many studies have indicated that with watershed areas, productivity growth has been mainly due to seed and fertilizer use. Thus, this state has to be given input subsidies for high yielding seed varieties but at the same time, the farmers need to be educated on the over use of chemical fertilizers.

With Bihar, agricultural performance is problematic on many fronts. First, although demographic pressure has increased and agricultural technology has improved, most of the uncultivated land is concentrated in southern Bihar, where irrigation facilities have not kept pace and the soil is of poor quality. Given the physiography of southern Bihar, wells are also unsuitable and thus the dominant mode of irrigation has been through tanks whose expansion and maintenance has been neglected. Second, the infrastructural facilities of Bihar have been lagging as seen by the infrastructure development. Due to infrastructural bottlenecks, availability of modern goods and services has not increased or their supply remains costly or unreliable.

Third, agriculture in Bihar is dominated by small and marginal farmers and the prevalence of mass poverty is largely related to the backwardness of agriculture. Fourth and importantly, the state agricultural policies in Bihar are in dire need of review. The semi-feudal production condition still exists in rural areas and the ineffective protection of tenancy rights has hindered agricultural growth. The slow pace of land consolidation reflects inadequate financial outlays and a shortage of manpower. Kalirajan and others (2001) note that marketing and extension services in Bihar are also rather weak compared to the other states.

Punjab on the other hand, was one of the few states which enjoyed the success of land reforms and the high priority of investment in rural infrastructure. Also, the irrigation base of the small and medium sized farms was comparable to that of large farms. In addition, the Punjab Agricultural University at Ludhiana contributed to the development of new seed varieties.

However, there are clear signs of a decline in crop yields since the 1990s and this has been associated with the increasing use of fertilizers and excessive water use which have increased the unit cost of production as a result of declining soil quality. Hence, care is needed when providing further input subsidies in fertilizer and water use. Another related fact is the steep increase in wages in Punjab and in the absence of productivity increases, the cost increase has affected the profitability of farmers. With Tamil Nadu, the main crop has been rice as this state is blessed with two monsoons. But from 1992-1997, there has been a steady decline in the areas irrigated by canals and an increase in well-irrigated areas while the use of tanks remains an unreliable source of irrigation. However, major improvements in about 10 rice varieties released in the early 1990s can be expected to improve productivity growth in rice production although pests and diseases as well as imbalance in the use of fertilizers are major constraints. Thus Tamil Nadu could do with subsidies of pesticides and farmers should be educated on the more effective use of fertilizers to obtain high yields. Interestingly, the cropping pattern of late has shown increasing substitution of food crops by commercial crops but there is concern that the benefits will reach farmers only with the development of adequate infrastructure such as roads and markets. However, that Tamil Nadu has a higher index than the all India average of infrastructure.

Challenges of Globalization

It is important to realize that globalization poses many challenges to a developing country like India, which had relied on a state directed and regulated policy regime for more than four decades. In moving to a more open, market-based economy there are many transitional problems that the country has to manage. The Government must play a pro-active role in facilitating the globalization process so that the opportunity sets for the economic agents are widened and the adverse effects of globalization are minimized. The Indian Government must also prepare the necessary information base and develop its capacity to articulate India's concerns and policy trade-offs in the international forums for multilateral trade and environmental negotiations.

In addition, the Government should embark on an extensive programme to educate farmers on the need to meet the standards required in the export markets. In fact, India needs to seek technical assistance in creating the capacity for meeting such standards and to consider watershed developments for environmental considerations. Equally important is the need to disseminate information about possible export markets to farmers, so that market access is achieved at minimum cost. Given the requisite information about markets and profitability, the likelihood of farmers investing in post-harvest and processing technologies and storage and efficient transportation arrangements as well as developing supporting infrastructure is very high.

Although the brave and bold move by India to reduce the tariff rate for agricultural products from 113 per cent in 1990-1991 to 26 per cent in 1997-1998 deserves to be applauded, the question of whether India is ready to compete in world markets remains to be seen. The infant industry argument may still hold for India to shield itself from external competition but one can easily question the length of time that is required to that end. Also, a delay in opening up to foreign trade has the danger that local producers may become too complacent and never be ready for competition.

As India opens up externally, it is also expected to face vulnerability in the wider international price fluctuations and thus Acharya (1998) claims that a minimum price support scheme is important. These prices can also act as a signal to adopt modern inputs and invest in yield-raising infrastructure for increasing production. For instance, keeping basic staple food grains at reasonable prices would induce farmers to switch over to high value crops. However, during the 1990s, Kalirajan and others (2001) shows that procurement prices especially for rice and wheat have been increasing faster than the general price level. Such high prices along with guaranteed purchases by the Food Corporation of India have pushed up market prices. These higher prices are partly responsible for the large buffer stocks with the Food Corporation. If this trend continues, India's comparative advantage will be eroded.

With openness and high price instability, unstable export revenue can also be expected. One way of reducing such risk is for India to diversify her agricultural exports. For example, since 1990, even in commodities such as tea, coffee, cocoa and spices, where India is supposed to have a comparative advantage international prices have been unstable. Besides increasing the type of exports to obtain more export revenue, India should also seriously consider exporting more value added agricultural products through agro-processing such as processed vegetables, fruits, fish and meat products given that export or even local demand for basic agricultural products would decline as incomes rise. The move to higher value added activities within the agricultural sector also spells greater opportunities for industrialization and vice versa as borne by Kalirajan and Shand's (1997) findings of a bi-directional relationship between agriculture and industry for most Indian states. On the other hand, Sivakumar and others (1999) establish empirical evidence of high forward linkages of agriculture due to the presence of agro-industries while Satyasai and Viswanathan (1999) show the significance of the spillover effects to the industrial sector via the intensive use of purchased inputs in the agricultural sector.

The lack or slow pace of internal or domestic liberalization is also seen to hinder the possible gains from external or trade liberalization. For example, although central zoning restrictions have been abolished, state government restrictions on inter-state and even inter-district restrictions on marketing and movement of goods still exist in many cases. This interferes with the benefits from crop specialization and economies of scale arising from comparative advantage. The land market is another example of distortion whereby land ceilings exist preventing the operation of large-sized farms. This has led to the emergence of a large number of small economically unviable land holdings. The easy leasing of land should be permitted with assurance of resumption. Yet another problem lies with the insufficiency of credit to agriculture. From 1995-1996, the Rural Infrastructure Development Fund was set up to allocate funds for the completion of projects and the government has committed itself to strengthening the cooperative credit structure through substantial refinancing and restructuring of the Regional Rural Banks. However, as mentioned earlier, due to varying institutional factors in the Indian states, these domestic reforms can be expected to yield quite different results.

Although India missed the opportunity to open up two decades ago, its attempts to do so now must be regarded as better late than never. Others such as Desai (1999) observe that, "the logic of the global economy as well as India's interests dictate that India become proactive in its liberalization policies. India must liberalize not because it has no choice but because it is the best choice".

His lament that India has adopted a 'victim mentality' when it really needs to adopt a 'winner mentality' has become less of a concern as over time, India has shown commitment to stay on the bandwagon of globalization. Having realized that globalization is a necessary but not a sufficient condition for high growth production, India has undertaken economic reforms, both internal and external. However, it must be ensured that these reforms are synchronized so that the pace of both reforms is set right in order to work hand in hand to promote agricultural productivity growth.

Thus, training the farmers and educating them appropriately to change their mindset and reorienting them to take up new activities or adopt foreign technology is of utmost importance. In this context, it is necessary to involve non-governmental organizations in training and mobilizing the rural poor to face the challenge of liberalization.

Also, with domestic economic reforms, more care needs to be exercised to draw up state-specific liberalization measures to maximize their benefits. Lastly, in the implementation of these reforms for successful globalization, one crucial element, not entirely within control is the need for good governance and stability in the political and economic environment. Political leaders who are the ultimate decision makers in these matters need to examine their own role dispassionately.

It is quite apparent that at this relatively early stage, there is little observable evidence of gains to India's agricultural performance after opening up. However, there could easily be benefits that have not yet surfaced, or are yet to be identified and perhaps too difficult or intangible to measure.

Whatever the case, it is highly likely that it is too soon to assess the full impact of globalization and economic reforms. Furthermore, the process of liberalization has been gradual and remains incomplete. For example, the complete removal of quantitative restrictions after March 2001 will have provided an opportunity for Indian farmers to tap world markets and, if they are successful, results should start to become evident soon. Export promotion via the development of export and trading houses as well as effective liberalizing export promotion zone schemes for agriculture are fairly recent measures and only time will tell as to how effective these measures are. Other possibilities such as agro-industry parks for promoting exports are also in the pipeline.

AGRICULTURAL DEVELOPMENT STRATEGIES

The lack of development in African countries has caused their economies to fall further and further behind those of the leading industrial nations. Many different development strategies have been tried. Some African countries have successfully encouraged investment in mining, tourism and industry. In agriculture, producers have been encouraged to move away from subsistence farming towards a more commercial approach as governments realised that income generated from the sale of surplus production could be used to improve productivity.

Agricultural development in ECA has faced an uphill struggle for the last twenty years. In an effort to stimulate development many countries borrowed heavily from bodies such as the IMF and from the commercial banking sector. These loans were not granted without strings attached, however. Most African countries were obliged to liberalise their economies by adopting significant policy changes often applied in packages known as Structural Adjustment Programmes (SAPs). These programmes included a number of elements but generally included requirement to:

- devalue the currency (to discourage imports and make exports more competitive),
- to make the currency freely convertible with other currencies,
- to cut public expenditure (in order to lower taxes),
- to dismantle state controlled marketing boards,
- to privatise state-owned industries (to raise capital and stimulate competition),
- to cut import restrictions (to encourage local industries to become more efficient),
- to allow foreign companies to freely repatriate profits (to encourage inward investment),
- and to boost exports.

The economists who designed SAPs were convinced that the only way African countries could transform their economies was to encourage inward investment and earn foreign exchange to invest in infrastructure and lay the foundations for industrialisation. These measures assumed that any country could compete in the world market if production and investment was concentrated in areas where they were deemed to have a competitive advantage. The only activity in which ECA nations could be said to have a competitive advantage in the world market was in the production of agricultural products and the exploitation of natural resources such as forestry, fishing and mining. The major flaw in this strategy was that similar advice was given to almost all tropical countries at the same time. Coffee-producing countries were encouraged to boost coffee production; sugar producers should produce more sugar, and so on. This resulted in over-production of these commodities which caused prices to plunge in the international markets. On average, current prices of tropical products (taking dollar inflation into account) are only about one seventh of those prevailing in 1980 (UN General Assembly). Economists call this phenomenon the fallacy of composition - less income is earned as more commodities are produced.

Another component of SAPs which many observers believe to have been counter-productive was the requirement to cut public expenditure. All too often this meant a cut in health programmes, education and agricultural extension. These measures have tended to reduce, rather than enhance the flexibility of the workforce and to curtail agricultural development. Overall, the record of inward investment has been poor and the ending of currency controls has increased opportunities for transfer pricing abuse (where companies over-price imports and under-price exports to reduce tax liability).

The most important SAP reform affecting the distribution of agricultural products has been the dismantling of state-controlled marketing boards and the practice of setting fixed purchasing and sales prices for commodities. It was assumed that government control of markets had obscured the forces of competition in supply and demand in the economy. A free market system would unleash these forces and increase productivity. It would force producers to meet the demands of consumers both in price and quality. Farmers would be able to buy inputs cheaper from competing suppliers, and the country, as a whole, would become more competitive in world markets.

Unfortunately, competitive and transparent markets did not emerge spontaneously (Shepherd). Most African farmers have too little land to produce truck-loads of goods and they are widely dispersed over the countryside. There is not enough business to encourage more than one trader to operate in many areas. Farmers have no means of communicating with the outside world or even the nearest town and they are often unwilling to risk the investment of bringing their goods to market resulting in considerable waste. Laws may have been passed which ban collusion among traders to pay low prices to farmers and charge high prices to consumers, but there are often insufficient resources to enforce such laws. Most traders have no experience of free market conditions and are reluctant to put their fellow traders out of business with serious competition.

Advocates of SAPs point to examples of countries that have improved their economies after adopting SAPs (World Bank) but there are few in Africa. Most ECA countries were not able to implement SAPs until relatively recently but rates of poverty have increased in many of these countries. Intense conflict, both within and between countries of the region, drought, desertification and, now HIV/AIDS have further weakened economic development in ECA. Most critics of the reform process, however, acknowledge that markets in African countries must be made more competitive and SAPs are designed to do that but this process may take a considerable time.

Trade Agreements

Economic links between ECA countries and their former colonial rulers have been maintained since independence. The economies of these countries have been moulded to meet the needs of their European counterpart for a hundred years or more and it would have been difficult for them to make the necessary changes in production patterns to trade successfully with other countries. The Europeans too needed to maintain supplies of raw materials and export markets in Africa and to protect the business of their trading companies.

In 1975 all ten countries covered by this study became party to the Lomé Convention. The Convention established trade, aid and cultural relationships between 15 European countries and 71 so called ACP (African, Caribbean and Pacific) countries which had either been colonies of, or had had strong historical links with, Europe. This agreement did not rule out bilateral or multilateral agreements with other countries but did give ACP countries preferential access to European markets. ECA countries have also decided to try to stimulate regional trade by bringing their economies closer together in regional economic agreements such as COMESA and SADC.

East Africans have exchanged goods and ideas with many other peoples of the world for millennia. In these exchanges of goods, cultural links have been established which have influenced ECA life at all levels – in religion, the arts, public sector structures, the economy and agriculture. In the last decade or two, however, this process has accelerated tremendously. There is no agreed definition of globalisation. It is simply a term which has been used recently to describe the impact of innovations in communication and transport systems on trade and the growing interdependence of nations due to economic sophistication and burgeoning output. In addition, high levels of protection between trading blocks of countries are breaking down as barriers to trade are reduced. These changes have made it possible to increase the volume of trade between countries in agricultural products.

It became clear that overall levels of trade could be increased if trade barriers were reduced, where there was agreement to do so, and that international trade should be governed by mutually agreed rules. The most active trading nations have been keen to find new markets for their goods and to reduce the barriers to free trade. These countries, however, have been reluctant to expose their own markets to foreign competition, especially unfair competition from subsidised or sub-standard goods.

At the international level, global liberalisation was stimulated by the General Agreement on Tariffs and Trade (GATT) which was first implemented in 1948 as a mechanism to promote free and fair trade among member countries. Several rounds of negotiations of trade rules have occurred throughout the history of GATT. The Uruguay Round, which began in 1986, was the eighth of the GATT rounds. In April 1994, officials from more than 100 countries gathered in Marrakech, Morocco to sign the Uruguay Agreement and to confer the role of further trade reforms on the newly established World Trade Organisation. The reform process is by no means complete. Almost all countries have now committed themselves to the objectives associated with their membership of the WTO. (6 out of the 10 countries covered in this study are WTO members). In order to meet these objectives, countries are obliged to further reform their existing internal economic and external trade policies. The future of trade and agriculture in ECA is inextricably linked to the rate and direction of these reforms.

AGRICULTURE AND VISION FOR THE FUTURE

Indian history can be broadly divided into five phases based on archeological findings:

1. Period of Saraswati (Harappan) civilization - 6500 B.C - 1000 B.C or also known as 'Vedic period' in history of India.
2. Golden period of Indian History - 500 B.C - 800 A.D
3. Muslim influence in India - 1000 A.D - 1700 A.D
4. British period in India - 1700 A.D - 1947 A.D
5. Modern India - 1947 - till date

The famine from 1876-78 led to institution of Famine Commission of 1880. George Nathaniel Curzon succeeded Lord Mayo as Viceroy of India. The horrors of Famine (1889-90) convinced Lord Curzon that urgent attention must be paid agriculture. Lord Curzon passed the Land Alienation Act (1900) and Cooperative Societies Act (1904). Lord Curzon, the Viceroy of India with the generous donations from Henry Phipps of the USA had founded the Imperial Agriculture Research Institute in 1905 at Pusa, a village in the Darabhanga district of Bihar. The main building at Pusa was named after its donor as the Phipps Laboratory. [PUSA stands for the donor of the Institute, Phipps of the USA].

There was a disastrous earthquake in 1936 and Pusa suffered heavily. After careful consideration the Government of India rebuilt the institute at New Delhi. The transfer to New Delhi was completed by October, 1936. The Marquees of Linlithgo, the then Viceroy of India, opened this Institutes in November, 1936.

This Institute (IARI) in Delhi is popularly known as the Pusa Institute. Under the University Grants Commission Act 1956, the Institute (at New Delhi) got the status of the Deemed University and Teaching and Research activities were intensified from 1958. In 1947, India had about 27 Agricultural and Veterinary Colleges including the Indian Agricultural Research Institutes, Indian Veterinary Research Institute and five other Agricultural Colleges established during the first decade of the century. These Colleges were purely teaching institutions affiliated to traditional universities and contributed little to research.

Agriculture Colleges were started at Poona (Pune) and Kanpur. Teaching was the main mandate. The Civil Veterinary Department was established in 1889, the main attention was on horse and male breeding. Systematic investigation on animal breeding started in 1890 with the Imperial Bacteriologist at the College of Science, Poona, Which was shifted to Mukteswer where the Imperial Bacteriologist Laboratory was established in1895.

This institution has been pioneer in the field of Veterinary research in the Country. Veterinary Colleges were started at Bombay, Lahore (now in Pakistan), Calcutta and Madras. The Indian Central Cotton Committee (ICCC) (1921) was formed as per recommendation of the Indian Central Cotton Commission (1917-18).

The Government of India appointed a Royal Commission in 1926 to examine the condition of agricultural and rural economy in India. The Imperial Council of Agricultural Research (ICAR) was established in 1929 as a Society under the Societies Registration Act, 1860. The Society was registered on July 16, 1929. [After Independence, the name of the society was changed to Indian Council of Agricultural Research (ICAR)]. The food crisis created by the

Second World War and the Bengal famine in 1943 deepened and become the matters of great concern to Government of India. To meet the food shortage the Grow More Food campaign was started in 1943.

On the recommendation of the Royal Commission on Agriculture in India (1927-1928), the Indian Lac Cess committee, through it had its origin in Lindsay-Harlow Committee (1921) got its statutory enactment in 1931 and the Indian Central Jute Committee (ICJC) on the line of cotton committee was set up in 1936. To undertake improvement and development of sugarcane, Jiggery (gur), Sugar and other bi-products the Indian Central Sugarcane Committee (ICSC) was constituted in 1944. The Development Council for sugar industries was formed in 1951. The ICSC was entrusted with responsibility of Research on Sugarcane. The development of Gur was entrusted to the All India Village Industries and Khadi Commission. The Indian Central Coconut Committee and the Indian Central Tobacco Committee were formed in 1945. The Indian Central Arecanut Committee was formed in 1949 and the Indian Central Spices and Cashewnut Committee were formed in 1958. Regional stations\sub-station on cotton, Jowar, Finger millet, setaria, castor, groundnut, linseed, bajra were established and the PIRRCOM (Project for Identification of Regional Research on Cotton, Oilseeds and Millets) were started.

Agricultural development has to be guided not only by compulsion of improving food and nutritional security but also by the concern for environmental protection, sustainability, profitability and even export. Crop productivity has to be improved in comparison with other countries. Further following the WTO agreement and liberalization process, the consequent liberalization process, the consequent globalization of markets would call for competitiveness and efficacy of agricultural production. The process of agricultural development could be accelerated and sustained only through investments on research and education.

All India Coordinated Research Projects: The AICRPs were born from the coordinated project on maize developed with the Rockefeller Foundation's assistance in 1957, ICAR has now about 70 All India Coordinated Research projects covering various disciplines and commodity crops, livestock, fisheries, home science, and agricultural engineering. An AICRP enables effective utilization of the resources in man and material anywhere in the country to tackle some of the important national problems.

The Indian Council of Agricultural Research is an autonomous apex body responsible for the: organization and management of research and education in all disciplines of agricultural sciences. It has been reorganized twice. In 1963 an expert committee (M.W.Parker Committee), was appointed by the

Government of India to inquire into the present set up and to suggest suitable changes in the ICAR. The Committee submitted its report in 1964. As per recommendations of the committee the ICAR become an autonomous body; its rules and by-laws were revised. A scientist heads as Director General (DG). To assist the DG, four posts of Deputy Director General (DDG)-Crop Science, Soils, Agronomy, Irrigation and Agricultural Engineering, Animal Sciences and Agricultural Education–were created. The Institute of Horticultural Research (Hessarghata), and Central Soil Salinity Research Institute (Karnal) were started. By this time there were 33 Research Institutes (25 in agriculture, 7 in Veterinary and animal husbandry and fishers and one in statistics) under the ICAR. In 1965, the ICAR became the nodal agency for coordinating agricultural research in the country.

It gained administrative control over the various institutes and commodity research institutes. Late Dr B.P.Pal took over as the first scientist as Vice President of ICAR. Dr. Pal instituted the All-India Coordinated Research Projects on various crops to integrate different disciplines and different institutions and Universities for an effective national grid of coordinated experiments.

He has been internationally acclaimed for this contribution In 1973, the AgrIcultural Research Service (ARS) was started by Dr M.S. Swaminathan, the first Director-General and Secretary of the Government of India and Dr Pal's, successor; to enable scientist's to move to other institutes within the system or sister organizations viz., the CSIR, BARC, etc. ICAR started the National Agricultural Research Project (Phase I) in 1983-94. NARP Phase II was wound up in 1992. Intensive Agricultural Area Programme (IAAP) was initiated in 1964. From 1966-67, High Yielding Variety Programme (HYVP) in crops like rice, wheat, maize, jowar, bajra, was started. The Krishi Vigyan Kendras (KVKs) and Trainers 'Training Centres (TTCs) were established on the recommendations of the Education Commission (1964-66).The Lab to Land Programme was launched by the ICAR in 1979 to extend and promote adoption of new technology amoung the small and marginal farmers and agricultural laborers to test the relevance of Technology under their socio-economic conditions.

ICAR Institutes: The ICAR is directly responsible for administering 32 research institutes in the fields of agriculture, animal sciences and fisheries. Some of these are single commodity-oriented crop institutions while a few of them undertake work on a number of crops.

The Indian Agricultural Research Institute (IARI), New Delhi, the Indian Veterinary Research Institute (IVRI), Izatnagar, and the National Dairy Research

Institute (NDRI), Karnal are the three national institutions which have responsibilities both for research and post-graduate education. Nevertheless, only the IARI has been given the status of a deemed university by virtue of which it awards its own post-graduate degrees in the field of agriculture. NDRI and IVRI are performing this function through affiliation with other universities. The recent establishment of the National Academy of Agricultural Management at Hyderabad as a constituent unit of the Council is an important landmark in institution building. This Academy would be responsible for providing quality training to various categories of personnel involved in agricultural research all over the country. Establishment of an Agricultural Research Service (ARS) started on October 1st, 1975 marks yet another landmark in the history of research management of ICAR.

Agricultural Universities: The responsibility for research in most of the States is now with the 21 agricultural universities, which perform in an integrated way the functions of teaching, research and extension education. The ICAR has recently taken major steps to further strengthen the agricultural research capabilities of the agricultural universities through the National Agricultural Research Project (NARP), which is being implemented through the assistance of IBRD.

Krishi Vigyan Kendras (KVKs): The ICAR has sponsored a programme known as the Krishi Vigyan Kendras, designed to provide skill oriented vocational training to practicing farmers, in-service field level extension Workers or those who intend to go in for self-employment.

VISION FOR AGRICULTURE IN 2020 AD

Every country needs a vision statement, which stirs the imagination and motivates all segments of society to a greater effort. It is an essential step in building a political consensus on a broad national development strategy, which encompasses, inter-alia, the roles and responsibilities of different agents in the economy, such as Central, State and local government, the private corporate sector, the small and tiny sector, people's organisation etc. It must identify potential risks and bottlenecks and their possible solutions in order to mobilise efforts in a focussed manner. It is clear, therefore, that to meet these objectives, a vision statement has to operate at several levels of generality and specificity. A vision is a picture of what is possible or what is desired in a longer-term future. It could be of one individual in origin or it could be a collective in its conception.

President APJ Abdul Kalam's address to the joint session of Parliament in 2003: The people to strive towards the goal of transforming India into a

Developed Nation by 2020. This vision captures our people's heightened self-confidence, rooted in India's impressive achievements in many fields. It also reflects the increased expectations of our people at the beginning of the new century, that India no longer be categorized as a developing, much less, a poor country. Nearly 260 million people, who are below the poverty line, want to join the mainstream of development. Our people are impatient to achieve 100 percent literacy, health for all, shelter for all, prosperity through knowledge-driven productivity, and a better quality of life—all of these enriched with our value system. Hence, it is time India launched a new vision, which I would call "Vision - 2020". To achieve this, they should concentrate on two mantras: Effective Implementation with People's Participation; and Effective Communication for People's Participation. A key element of "Vision 2020" would be "*Providing Urban amenities in Rural Areas* (PURA)". More than two-thirds of India's population lives in rural areas. We need to give a new thrust to their all-round development through a mega mission for their empowerment.

The richness and diversity of India's bio-resources are a major gift of nature to us. The Biological Diversity Bill 2002, passed in the Winter Session, marked a major milestone in India's commitment to conservation and sustainable utilization of our bio-resources. An ambitious afforestation programme with people's participation that establishes Joint Forest Management Committees in all the 1.73 lakh villages situated on the fringes of the forest areas has been launched.

The scope of the National River Conservation Plan has been considerably broadened to include works in 155 towns along polluted stretches of 29 rivers spread over 17 States. India successfully hosted the Eighth Conference of Parties to the United Nations Framework Convention on Climate Change in New Delhi last year. The successful adoption of the Delhi Declaration helped to raise awareness of developing country concerns in climate change. India welcomes the adoption of the Plan of Action at the World Summit on Sustainable Development, which was held in Johannesburg last year. India's first meteorological satellite was successfully launched. The forthcoming launches of satellites in the INSAT-3 series will add further capacity to the INSAT system, which is already one of the largest domestic communication satellite systems in Asia. An exclusive satellite for education, EDUSAT, is also under development. ISRO has taken up the task of tele-medicine connectivity to provide medical services to remote areas. The Indian Remote Sensing Satellites continue to provide valuable data for our resources survey and management. Ground water prospect maps for six States were released recently to help locate sites for drilling borewells.

The Nation has been searching for a lasting solution to the recurring problem of droughts and floods, which have been taking a huge human and economic toll.

Networking of our river systems to transfer water from the surplus basins to the areas of deficit has engaged people's attention for many decades. The Government has set up a Task Force to prepare a practical blueprint for this project, without compromising environmental safety and the interest of displaced people.

This initiative will bring significant benefits in drinking water, irrigation, power generation, inland navigation, and tourism. I must emphasize that this mega project does not negate the need for promoting small and micro programmes for water conservation at local levels. The two are mutually complementary. The National Water Resources Council has adopted a new National Water Policy emphasizing integrated water resources development and management for optimal and sustainable utilization of available surface and ground water. The Centre has launched a Fast Track Programme for the completion of those major and medium irrigation schemes that can be completed in one year.

Subsequent to the approval by the Narmada Control Authority, the dam height was raised, and this has mitigated the problem of drinking water and irrigation in arid areas of Saurashtra and North Gujarat. The policy of procurement at the Minimum Support Price, while ensuring remunerative prices for wheat and rice farmers in surplus States, has resulted in huge stocks of rice and wheat with the public agencies. As a response to this, the Government has been encouraging exports of food grains. The wide-ranging recommendations on long-term food management made by the High Level Committee are being examined.

There is an urgent need to review the current policies, which have impeded crop diversification and led to unsustainable food subsidies, and to ensure crop neutral support to our farmers without excessive procurement. Fertilizers are a critical component in our scheme of food security. The new pricing policy for urea to be implemented from April 2003 aims at greater transparency, efficiency, and fiscal discipline. While the Government is committed to deregulate the marketing and distribution of fertilizers, it would ensure that major fertilizers are available in the country both in adequate quantity and quality at affordable prices to farmers in all the States. The sugar industry has lately faced serious difficulties, constraining the capacity of sugar factories to make timely payment to sugarcane farmers.

Several steps have been initiated to protect the interests of sugarcane growers, while ensuring viability of sugar mills. Sustained efforts are being made to promote horticulture as a major area of diversification in agriculture. The cold storage scheme is working well and has created an additional capacity of 28 lakh tonnes.

A new scheme of construction, renovation, and expansion of rural godowns known as Grameen Bhandaran Yojana has been launched. This scheme will help prevent distress sales by small and marginal farmers. A new National Policy on Cooperatives has been announced. A National Seeds Policy has been finalized. Under the scheme of Agriclinics and Agribusiness Centres, launched last year, unemployed agriculture graduates provide extension services to the farmers on payment. Recognizing the need for value-addition in agricultural and horticultural produce, the Government has given high priority to the development of food processing industries. A Group of Ministers has been constituted to propose a single modern integrated food law and related regulations, to replace the existing myriad laws, which have affected the growth of this sector.

6

Crop Diversification in Small Farming Systems

AUTOMATION OF SMALL FARMS

It is a mistaken notion that it is neither possible nor desirable to mechanise small farms. Japan, a more densely populated country, with average landholding no bigger than ours, has fully mechanised its agriculture with great success. Before mechanisation in Japan, it used to take 1600 to 2000 manhours to produce 4.7 tonnes of rice on one hectare of land. After mechanisation, it takes only one third of it to produce 5.2 tonnes of rice on the same one hectare of land. Even in India, an energy survey in Punjab has revealed that human and bullock energy is several times costlier than energy produced by a diesel engine or an electric motor.

Production cost of farm products cannot be reduced except through mechanisation, which does not necessarily require the use of big machines. For small holdings, small machines are available at reasonable prices all over the world, except in India.

In order to minimise the drudgery of small farmers, increase the efficiency of inputs-use and save farmers' time for enabling them to take up income-augmenting supplementary enterprises such as dairying, goat keeping and sericulture, the use of modern time-saving farm implements of appropriate size needs to be promoted, either by duty-free import of such implements from countries like Japan and South Korea or bysubsidising their purchase for small farmers. The use of small mechanised tools by small farmers who do not exploit labour, does not reduce employment, but only adds value to the working hours and that is exactly what is needed to lift them above the poverty level.

Need for a Third Marketing Channel

The greatest hurdles in the path of progress of farmers are the two prevailing marketing systems-one private and the other public both of which are exploiting the producers as well as the consumers. Therefore, there is need for a third marketing channel, through either revitalisation of the existing co-operatives or empowerment of the Gram Panchayats to act as co-operatives.

FARM LEVEL DIVERSIFICATION IN DRYLAND REGIONS

Plot diversification arises from farmers owning more than one field, leasing in land, sub-ploting within fields, and multiple cropping over cropping season. More land fragmentation implies greater plot diversification. Since, majority of farmers in dryland regions are small and marginal farmers, they have to meet multiple objectives with limited land. Hence, they plan their cropping activities in such a manner that it can ensure them regular income, though low, and minimise risk of crop failure. Land fragmentation has many inherent limitations as there is a feeling that it leads to slow adoption of new technology. But for farmers especially marginal and small, it is risk management strategy for spreading risk to larger areas. Some a studies have been conducted based on country or district level data using gross revenue variability as a measure of crop income instability but they often fail to capture real farm level realities. For example, the study by Barah and Binswanger (1982) using district level data for four states indicates that crop diversification was largely a response to low and unstable rainfall. Moreover crop diversification does significantly reduce gross revenue variability. The effect of diversification has not been adequately addressed with farm level data on crop income stability.

Data

The present study uses farm level data from three distinct agroclimatic regions covering a period of five years to address the causes and consequences of inter-regional and inter-village differences in diversification behaviour. This data set from the ICRISAT Village Studies pertain to a panel of 30 cultivator households in each of the six villages located in three broad soil, climatic and cropping regions of the semi-arid tropics of India.

DETERMINANTS OF CROP DIVERSIFICATION

The level of crop diversification, both within and across villages, largely depends on site specific ecological characteristics, the resource endowment of farm households and their personal characteristics, emerging weather conditions, crop rotational requirements, perceived consumption needs of the households and market access. However, all these factors such as crop rotational thoughts and consumption targets were not readily available with the data and so they have not been analysed.

It was expected that personal characteristics, including risk aversion lead to greater diversification. Risk attitudes for each household were estimated by experimental games described in Binswanger (1978) and Binswanger et al (1982). Changing weather conditions particularly at planting is expected to influence crop choices and levels of diversification. The effect should be stronger in less assured rainfall regions where crops are planted in post-rainy season on residual soil moisture.

The resource endowments of the farm household is expected to have a substantial influence on diversification.

The access to irrigation, availability of draft power, family labour, land quality, experience and education of the farmers, etc, play an important role in influencing diversification. Since these three regions have diverse agroclimatic conditions the determinants of farm-level diversification are analysed separately for each region. A regional analysis is supported by F test that rejects at the 5% level, the hypothesis that the estimated regressions coefficients are equal across the three regions.

The regression results indicate that draft power availability is an important explanatory of variation in crop diversification across households within some of the regions. As expected, larger farms with more gross cropped area are more diversified than their smaller counterparts. In contrast, inter household differences in family size have little bearing on variation in the levels of diversification. In Sholpur region, resource endowment thoughts are less important in explaining diversification which is largely accounted for by differences in land quality, cropping-year conditions, and village disparities within the region.

The influence of irrigated area is also significant and site specific. Limited well irrigation, particularly in the post-rainy season opens up opportunities to grow more crops such as wheat, chickpea, and other pulses.

Differences in risk aversion are strongly and positively associated with crop diversification in one region but in other regions do not play a strong role in determining crop diversification. Cropping year and site specific village characteristic influences have a negligible impact on crop diversification in the same region. However, in drought prone area of Sholapur crop diversification appears to be sensitive to changing weather conditions.

CHANGE AND HOUSEHOLD NET CROP INCOME

It is almost taken for granted in the risk literature that increasing diversification beyond a certain level leads to reduced risk at the cost of expected net crop income.

The size of such a trade-off is an empirical question which is addressed in this section. Household net crop income is defined as returns to land, capital and management, i.e. the cost of all purchased inputs and the value of family labour and owned draft power are deducted from the value of production. Household net crop income averaged Rs. 2,150, 2,190, 2,370 over the five cropping years for Mahbubnagar, Sholapur, and Akola regions respectively.

Household net cropping-year income is regressed on crop and plot diversification, land quality (in the Sholapur region), irrigated area, land value, gross cropped area, age and education. Cropping-year dummies and within-region village variables are included to capture the average effects of changes in net crop income over time and between villages in the same region. The diversification indices, GDI and PDI, are also squared and included in the regression to capture possible non-linear effects of diversification on income. Because the effect of the resource base on farm level diversification and ultimately on crop income is sequential, the inclusion of resource base determinants and levels of diversification as independent variables should not lead to simultaneity bias with ordinary least squares estimation. Nonetheless, it is better to draw inferences of association rather than causality from the estimated coefficients.

The importance of land quantity and quality in determining net crop income. In the Mahbubnagar villages, the significant quadratic response between crop diversification and income suggests that substitution away from the dominant crops particularly paddy does result in lost income. In Sholapur and Akola crop diversification does not appear to have reached a level that results in reduced income.

With the possible exception of Akola, the results suggest that fragmentation has not yet reached a level where inefficiencies associated with scattered holdings and small plot size noticeably detract from crop income. The significant non-linear effects of plot diversification on income in Mahbubnagar are puzzling but may be explained by some complicated interactions between multiple cropping which is positively associated with plot diversification and irrigation. These effects may not have been fully captured by the irrigated area variable.

AGROECOLOGY, SMALL FARMS AND FOOD REIGN

Universal forces are challenging the ability of developing countries to feed themselves. A number of countries have organized their economies around a competitive export-oriented agricultural sector, based mainly on monocultures. It may be argued that agricultural exports of crops such as soybeans from Brazil make significant contributions to the national economies by bringing in hard currency that can be used to purchase other goods from abroad. However, this type of industrial agriculture also brings a variety of economic, environmental, and social problems, including negative impacts on public health, ecosystem integrity, food quality, and in many cases disruption of traditional rural livelihoods, while accelerating indebtedness among thousands of farmers.

The rising push toward industrial agriculture and globalization—with an emphasis on export crops, lately transgenic crops, and with the rapid expansion of biofuel crops (sugar cane, maize, soybean, oil palm, euca-lyptus, etc.)—is increasingly reshaping the world's agriculture and food supply, with potentially severe economic, social, and ecological impacts and risks. Such reshaping is occurring in the midst of a changing climate expected to have large and far-reaching effects on crop productivity pre-dominantly in tropical zones of the developing world. Hazards include increased flooding in low-lying areas, greater frequency and severity of droughts in semiarid areas, and excessive heat conditions, all of which can limit agricultural productivity.

Globally, the Green Revolution, while ornamental crop production, proved to be unsustainable as it damaged the environment, caused dramatic loss of biodiversity and associated traditional knowledge, favored wealthier farmers, and left many poor farmers deeper in debt. The new Green Revolution proposed for Africa via the multi-institutional Alliance for a Green Revolution in Africa (AGRA) appears destined to repeat the tragic record left by the fertilizer dependent miracle seeds, in Latin America and Asia by increasing dependency on foreign inputs and patent-protected plant varieties which poor farmers cannot afford (for example, fertilizer costs went up approximately 270 percent last year) and on foreign aid.

In the face of such global trends, the concepts of food sovereignty and ecologically based production systems have gained much attention in the last two decades. New approaches and technologies involving application of blended modern agroecological science and indigenous knowledge systems spearheaded by thousands of farmers, NGOs, and some government and academic institutions have been shown to enhance food security while conserving natural resources, biodiversity, and soil and water throughout hundreds of rural communities in several regions. The science of agroecology—the application of ecological concepts and principles to the design and management of sustainable agricultural ecosystems—provides a framework to assess the complexity of agroecosystems. This approach is based on enhancing the habitat both above ground and in the soil to produce strong and healthy plants by promoting beneficial organisms while adversely affecting crop pests (weeds, insects, diseases, and nematodes).

For centuries the agricultures of developing countries were built upon the local resources of land, water, and other resources, as well as local varieties and indigenous knowledge. This has nurtured biologically and genetically diverse smallholder farms with a robustness and a built-in resilience that has helped them to adjust to rapidly changing climates, pests, and diseases. The persistence of millions of agricultural hectares under ancient, traditional management in the form of raised fields, terraces, polycultures (with a number of crops growing in the same field), agroforestry systems, etc., document a successful indigenous agricultural strategy and constitutes a tribute to the "creativity" of traditional farmers. These microcosms of traditional agriculture offer promising models for other areas because they promote biodiversity, thrive without agrochemicals, and sustain year-round yields. The new models of agriculture that humanity will need to include forms of farming that are more ecological, biodiverse, local, sustainable, and socially just. They will be rooted in the ecological rationale of traditional small-scale agriculture, representing long established examples of successful community-based local agriculture. Such systems have fed much of the world for centuries and continue to feed people in many parts of the planet.

Fortunately, thousands of small customary farms still exist in most rural landscapes of the third world. The productivity and sustainability of such agroecosystems can be optimized with agroecological approaches and thus they can form the basis of food sovereignty, defined as the right of each nation or region to maintain and develop their capacity to produce basic food crops with the corresponding productive and cultural diversity. The emerging concept of food sovereignty emphasizes farmers' access to land, seeds, and water while focusing on local autonomy, local markets, local production-consumption cycles, energy and technological sovereignty, and farmer-to-farmer networks.

SMALL FARMERS AND REGIONAL FOOD SECURITY

In Latin America, there were about 16 million peasant production units in the late 1980s, occupying close to 60.5 million hectares—34.5 percent of the total cultivated land. The peasant population includes 75 million people representing almost two-thirds of Latin America's total rural population. The average farm size of these units is about 1.8 hectares, although the contribution of peasant agriculture to the general food supply in the region is significant. These small units of production were responsible for 41 percent of the agricultural output for domestic consumption and for producing at the regional level 51 percent of the maize, 77 percent of the beans, and 61 percent of the potatoes. The contribution to food security of this small-farm sector is today as crucial as twenty-five years ago.

Africa has approximately 33 million small farms, representing 80 percent of all farms in the region. The majority of African farmers (many of them are women) are smallholders, with two-thirds of all farms below 2 hectares and 90 percent of farms below 10 hectares. Most small farmers practice "low-resource" agriculture which is based primarily on the use of local resources, but which may make modest use of external inputs. Low-resource agriculture produces the majority of grains, almost all root, tuber, and plantain crops, and the majority of legumes. Most basic food crops are grown by small farmers with virtually no or little use of fertilizers and improved seed. This situation, however, has changed in the last two decades as food production per capita has declined in Africa. Once self-sufficient in cereals, Africa now has to import millions of tons to fill the gap. Despite this increase in imports, smallholders still produce most of Africa's food.

In Asia, China alone accounts for almost half the world's small farms (on 193 million hectares), followed by India with 23 percent, and Indonesia, Bangladesh, and Vietnam. Of the majority of more than 200 million rice farmers who live in Asia, few cultivate more than 2 hectares of rice. China has probably 75 million rice farmers who still practice methods similar to those used more than 1,000 years ago. Local cultivars, grown mostly on upland ecosystems and/or under rain-fed conditions, make up the bulk of the rice produced by Asian small farmers.

Represent a Refuge of Agrobiodiversity Free of GMOs

Traditional small-scale farmers tend to grow a wide variety of cultivars. Many of these plants are landraces, more genetically heterogeneous than formal modern varieties, and grown from seed passed down from generation to generation. These landraces offer greater defences against vulnerability and enhance harvest security in the midst of diseases, pests, droughts, and other stresses. In a worldwide survey of crop varietal diversity on farms involving twenty-seven crops, scientists found that considerable crop genetic diversity continues to be maintained on farms in the from of traditional crop varieties, especially of major staple crops. In most cases, farmers maintain diversity as insurance to meet future environmental change or social and economic needs. Many researchers have concluded that variety richness enhances productivity and reduces yield variability. Given the penetration of transgenic crops into centres of diversity, at issue is the possibility that traits important to indigenous farmers (resistance to drought, competitive ability, performance in polycrop systems, storage quality, etc.) could be traded for transgenic qualities (e.g., herbicide resistance) which are of no importance to farmers that do not use agrochemicals. Under this scenario, risk will increase and farmers will lose their ability to produce relatively stable yields with a minimum of external inputs under changing environments. The social impacts of local crop shortfalls, resulting from changes in the genetic integrity of local varieties due to genetic pollution, can be considerable in the margins of the developing world.

It is crucial to protect areas of peasant agriculture free of contamination from GMO crops. Maintaining pools of genetic diversity, geographically isolated from any possibility of cross fertilization or genetic pollution from uniform transgenic crops, will create "islands" of intact genetic resources to act as safeguards against the potential ecological failure derived from the Second Green Revolution increasingly being imposed with programs such as the Gates-Rockefeller AGRA in Africa. These genetic sanctuary islands will also serve as the only source of GMO-free seeds that will be needed to repopulate the organic farms in the North that will inevitably be contaminated by the advance of transgenic agriculture. The small farmers and indigenous communities of the Global South, with the help of scientists and NGOs, can continue being the creators and guardians of a biological and genetic diversity that has enriched the food culture of the whole planet.

Productive and Reserve Conserving

Although the conventional wisdom is that small family farms are backward and unproductive, research shows that small farms are much more productive than large farms if total output is considered rather than yield from a single crop. Maize yields in traditional Mexican and Guatemalan cropping systems are about 2 tons per hectare or about 4,320,692 calories, sufficient to cover the annual food needs of a typical family of 5-7 people. In the 1950s the*chinampas* of Mexico (raised growing beds in shallow lakes or swamps) had maize yields of 3.5-6.3 tons per hectare. At that time, these were the highest long-term yields achieved anywhere in Mexico. In comparison, average maize yields in the United States in 1955 were 2.6 tons per hectare, and did not pass the 4 tons per hectare mark until 1965.10 Each hectare of remaining chinampa can still produce enough food for 15-20 persons per year at a modern subsistence level.

Traditional manifold cropping systems provide as much as 20 percent of the world food supply. Polycultures constitute at least 80 percent of the cultivated area of West Africa, while much of the production of staple crops in the Latin American tropics also occurs in polycultures. These diversified farming systems in which the small-scale farmer produces grains, fruits, vegetables, fodder, and animal products in the same field or garden out-produce the yield per unit of single crops such as corn grown alone on large-scale farms. A large farm may produce more corn per hectare than a small farm in which the corn is grown as part of a polyculture that also includes beans, squash, potatoes, and fodder. But, productivity in terms of harvestable products per unit area of polycultures developed by smallholders is higher than under a single crop with the same level of management. Yield advantages can range from 20 percent to 60 percent, because polycultures reduce losses due to weeds (by occupying space that weeds might otherwise occupy), insects, and diseases (because of the presence of multiple species), and make more efficient use of the available resources of water, light, and nutrients.

By managing fewer resources more intensively, small farmers are able to make more profit per unit of output, and thus, make more total profits—even if production of each commodity is less. In overall output, the diversified farm produces much more food. In the United States the smallest two-hectare farms produced $15,104 per hectare and netted about $2,902 per hectare. The largest farms, averaging 15,581 hectares, yielded $249 per hectare and netted about $52 per hectare. Not only do small-to medium-sized farms exhibit higher yields than conventional larger-scale farms, but they do this with much lower negative impacts on the environment, as research shows that small farmers take better care of natural resources, including reducing soil erosion and conserving biodiversity. However, an important part of the higher per hectare income of small farms in the United States is that they tend to by-pass middlemen and sell directly to the public, restaurants, or markets. They also tend to receive a premium for their local, and frequently organic, products.

The inverse relationship between farm size and output can be attributed to the more efficient use of land, water, biodiversity, and other agricultural resources by small farmers. So in terms of converting inputs into outputs, society would be better off with small-scale farmers. Building strong rural economies in the Global South based on productive small-scale farming will allow the people of the South to remain with their families in the countryside.

This will help to stem the tide of out-migration into the slums of cities that do not have sufficient employment opportunities. As the world's population continues to grow, redistributing farmland may become central to feeding the planet, especially when large-scale agriculture devotes itself to feeding cars through growing agrofuel feedstocks.

COSMETIC THE PRODUCTIVITY OF SMALL FARMING SYSTEMS

Despite the evidence of the resiliency and productivity advantages of small-scale and traditional farming systems, many scientists and growth specialists and organizations argue that the performance of subsistence agriculture is unsatisfactory, and that agrochemical and transgenic intensification of production is essential for the transition from subsistence to commercial production. Although such intensification approaches have met with much failure, research indicates that traditional crop and animal combinations can often be adapted to increase productivity. This is the case when ecological principles are used in the redesign of small farms, enhancing the habitat so that it promotes healthy plant growth, stresses pests, and encourages beneficial organisms while using labour and local resources more efficiently.

Several reviews have amply documented that small farmers can produce much of the needed food for rural and neighbouring urban communities in the midst of climate change and burgeoning energy costs. The evidence is conclusive: new agroecological approaches and technologies spearheaded by farmers, NGOs, and some local governments around the world are already making a sufficient contribution to food security at the household, national, and regional levels. A variety of agroecological and participatory approaches in many countries show very positive outcomes even under adverse environmental conditions. Potentials include: raising cereal yields from 50 to 200 percent, increasing stability of production through diversification, improving diets and income, and contributing to national food security (and even to exports) and conservation of the natural resource base and biodiversity. This evidence has been reinforced by a recent report of the United Nations Conference on Trade and Growth stating that organic agriculture could boost African food security. Based on an analysis of 114 cases in Africa, the report revealed that a conversion of farms to organic or near-organic production methods increased agricultural productivity by 116 percent.

Moreover, a shift towards organic production systems has enduring impact, as it builds up levels of natural, human, social, financial, and physical capital in farming communities. The International Assessment of Agricultural Knowledge, Science and Technology (AKST) commissioned by World Bank and the Food and Agriculture Organization (FAO) of the United Nations recommended that an increase and strengthening of AKST towards agroecological sciences will contribute to addressing environmental issues while maintaining and increasing productivity. The assessment also stresses that traditional and local knowledge systems enhance agricultural soil quality and biodiversity as well as nutrient, pest, and water management, and the capacity to respond to environmental stresses such as climate.

Whether the potential and spread of agroecological innovations is realized depends on several factors and major changes in policies, institutions, and research and growth approaches. Proposed agroecological strategies need to target the poor deliberately, and not only aim at increasing production and conserving natural resources. But they must also create employment and provide access to local inputs and local markets. Any serious attempt at developing sustainable agricultural technologies must bring to bear local knowledge and skills on the research process. Particular emphasis must be given to involving farmers directly in the formulation of the research agenda and on their active participation in the process of technological innovation and dissemination through Campesino a Campesino models that focus on sharing experiences, intensification local research, and problem-solving capacities. The agroecological process requires participation and enhancement of the farmer's ecological literacy about their farms and resources, laying the foundation for empowerment and continuous innovation by rural communities.

Equitable market opportunities must also be developed, emphasizing local commercialization and distribution schemes, fair prices, and other mechanisms that link farmers more directly and with greater solidarity to the rest of the population. The ultimate challenge is to increase investment and research in agroecology and scale up projects that have already proven successful to thousands farmers. This will generate a meaningful impact on the income, food security, and environmental well-being of all the population, especially small farmers who have been adversely impacted by conventional modern agricultural policy, technology, and the penetration of multinational agribusiness deep into the third world.

CHANGE OF SMALL FARMS

Diversification of small farms is often suggested as a means for rapid rural growth in India. Small and marginal holdings account for about three-fourth of the total operational holdings in the country, operating over one-fourth of the total area. Majority of small and marginal farmers cultivate mainly low value, subsistence crops. In the absence of adequate farm and non-farm employment opportunities, they are also forced to live below poverty line. The situation is likely to worsen because of the growing pressure of population on land and the limited scope of increasing additional production through subsistence farming. Hence arises the need for commercialisation and diversification of small farms within and outside agriculture and their proper integration with local and global markets. This is intended not only to liberate the small and marginal farmers from the poverty trap, but also to meet the country's growing demands for fruits, vegetables, milk and milk products, meat, fish, eggs etc. which generally show rising trends with increasing levels of per-capita income in the economy. However, the moot question today is how do we achieve the goal of diversification of small farms by overcoming various technological, infrastructural, institutional and policy constraints. The present paper intends to analyse the problems and prospects of diversification of small farms in India, mainly by keeping in view some of the above mentioned constraints. It also raises issues that need our attention on priority basis for accelerating the pace of diversification of small farms.

RATIONALE FOR SMALL FARM DIVERSIFICATION

By small farm diversification we mean that due to economic pressure and incentives, small landholders would (i) shift from subsistence farming to commercial farming, (ii) shift from low value food/non-food crops to high value food/non-food crops and switch over from local to high yielding plant varieties. The diversification would also mean that small farms would not only undertake seasonal crop farming, but also animal husbandry, fishing, agro-forestry, horticulture etc. and would participate in industrial and other non-farm economic activities as either self-employed or wage earners, for supplementing their incomes.

The average size of holdings of small and marginal farmers in India is so low that it cannot generate adequate employment and income for their sustenance, based on crop cultivation only. This is particularly true if the small farmers use only traditional technology having local, low yielding plant varieties and low value subsistence crops and if they do not undertake other farm and non-farm activities including horticulture, sericulture, animal husbandry, fishing, agro-forestry, agro-processing, agro-services etc.

It is true that specialisation in any particular crop farming or any other enterprise may enable the farmers to develop managerial expertise and also reduce the per unit cost of production. But specialised farming may not provide them the potential benefits of a synergistic production system, involving a tight knit integration of all possible economic activities, such that the net aggregate output is greater than the sum of the system's individual parts (Taylor, 1994). In other words, farmers can take advantage of production complementarities and supplementarities and thereby reap the benefits of synergism through appropriate choice of crop combinations or other economic activities. This also enables the farmers to maximize efficiency of resource use, through (i) multi dimensional use of limited land, (ii) multi dimensional use of time, (iii) enterprise symbiosis and (iv) use of products of one enterprise as inputs to another enterprise on the same farm and (v) intensive use of fixed family labour, involving integration of farm and non-farm activities.

Besides, ceteris paribus, diversified farm generally have higher growth and stability of farm income, as the risks due to fluctuating production and market prices would be minimum, as all crops/enterprises on the same farm would not face adverse weather condition, pest/insect attack or price uncertainties the same season or year. In addition, unlike specialised monoculture which can experience additional pest problems and soil erosion, diversified farming provides eco-system check and balance mechanism for controlling pests. For instance, crop rotations and relay cropping provide natural defences against weeds, insects and diseases and also enable the farmers to make full use of limited land and time. Also, diversified crop-livestock farms can make more efficient use of forages, crop residues and other low value components of crop rotations and enjoy the benefits of manure, the application of which can build up soil nutrients, organic matter, tilth and other dimensions of soil fertility.

Nature and Extent of Agricultural

The available data show that about 32 percent of India's Gross Domestic Product (GDP) originate from the agricultural sector and nearly 66 percent of the total labour force are primarily engaged in agriculture. Probably India is one of the very few Asian countries with such high share of agricultural sector in GDP as well as total employment. This does not compare well with a country like Japan where agriculture contributes hardly 2 percent of GDP and employs only 6 percent of the total labour force or a country like South Korea where the shares of agriculture in total GDP and employment are only 8 per-cent and 23 percent respectively. In India, the percentage share of agriculture in GDP declined from 47 percent in 1965 to 32 percent in 1992 and that of agricultural labour force to total labour force decreased only marginally from 73 percent in 1965 to 66 percent in 1992. In most other countries, the pace of non-farm diversification has been much more rapid, thereby enabling the growing labour force to be absorbed outside agriculture. The trends in horizontal agricultural diversification in selected Asian countries. Even within agriculture, the pace of diversification in favour of non-cereal crop enterprises in India appears to be quite modest, as compared to countries like South Korea, China and Indonesia where the growth rates of production of fruits, vegetables, meat, milk and fish are phenomenal. In Thailand also, the growth rates of production of milk and fish as well as agro-industrial output are faster than those in India.

The pace of agricultural diversification in India varied widely from region to region. Although crop production accounts for the lion's share in net domestic product of the agricultural sector in all the states, in recent years, there has been a phenomenal growth in the contributions of animal husbandry and poultry to the net aggregate domestic product of the agricultural sector in states like Andhra Pradesh, Bihar, Gujarat, Himachal Pradesh, ' Karnataka, Kerala, Madhya Pradesh (M.P.), Rajasthan and Uttar Pradesh. The percentage share of fishing also increased significantly in several states including Bihar, Gujarat and West Bengal. Considering the country as a whole, the animal husbandry and fishing contributed about 15.8 percent and 2.1 percent to GDP respectively. The share of agro-industries in total GDP of the rural sector increased from about 3.7 percent in 1970-71 to 5.3 percent in 1988-89.

In recent years, there has been renewed emphasis on promotion of diversified agricultural growth in favour of horticultural and vegetable crops. The production of fruits in the country increased from 20 million tonnes in 1980 to 32 million tonnes in 1993. The production of mango increased from 8 million tonnes in 1980 to 10 million tonnes in 1993 and that of bananas rose from 4 million tonnes in 1980 to 7 million tonnes in 1993. The production of oranges and pears almost doubled during this period, while that of grapes and papaya increased by three times. The production of tomato crop increased from 1.5 million tonnes in 1980 to 4.6 million tonnes in 1993 and that of dry onions rose from 2.5 million tonnes in 1980 to 4 million tonnes in 1993. However, the production of other vegetable crops did not increase significantly.

The proportion of area under food grains declined from 75.6 percent in 1960-61 to 69.8 percent in 1989-90 mainly due to the declining area under crops like jowar, bajra, barley, millets and pulses, there has been a steady upward trend in the growth of area under oilseeds and fruits and vegetables. The area under fruits and vegetables increased from about 2.6 million hectares in 1960-61 to 6.1 million hectares in 1989-90. However, area under rice and wheat among cereal crops maintained an upward trend due to relative yield and economic advantages of growing HYV rice and wheat. The area under tomato increased from about 50,000 hectares in 1960-61 to 3,10,000 hectares in 1993. In fact, wherever irrigation facilities are available, crops like rice, wheat, sugarcane and fruits and vegetables find favour with farmers, irrespective of their size of holdings.

Need for National Level Agricultural Diversification

From national or social point of view, diversified farming can be helpful in protecting the agricultural environment from the loss of natural 'bio-diversity' for sustainable growth. A country which produces and markets only a few specialised crops or crop based enterprises is more vulnerable to low income risk due to fluctuation in domestic and international prices or whenever a natural disaster affects production of the specialised farming system. Thus, for the purpose of national food security, price/income stability and protection of bio-diversity, there is need for maintaining a tender balance between those farming activities which are potentially more profitable and those which are relatively less profitable, but eco friendly. On the whole, it is generally agreed that diversified farming system if adopted in a planned manner may lead to greater individual as well as social gains. It is in the interest of both individual small farmers and the nation as a whole to plan for horizontal diversification within agriculture as well as vertical diversification, involving income earning activities by small farms outside agriculture.

EXTENT OF SMALL FARM DIVERSIFICATION

The agricultural census data as well as various micro level studies conducted in different regions indicate that small and marginal farms adopt diversified farming, despite very limited space and low sizes of holdings. However, the extent of diversification is relatively low on small and marginal farms, as compared to large farms. The proportion of area under non-food crops ranged between 11.4 percent in the case of marginal farms below 1 hectare to 26.8 percent in the case of large farms above 10 hectares. Further, scheduled caste and scheduled tribe farmers among all marginal and small farmers allocated relatively higher proportion of their tiny plots to non-food crops.

Besides, the results of a number of micro level studies conducted in different regions of the country indicate that small and marginal farms are not averse to diversification. A study on Diversification of Punjab Agriculture by A. J. Singh et al (IJAE, 1985) shows that there is an inverse relationship between farm size and agricultural diversification. Gupta and Tiwari (IJAE,

1985) analysed the data of Allahabad district in Uttar Pradesh and observed that small farms are relatively more diversified. Thakur, Kapila and Moorti (1985) in their study on Vegetable Production for Diversification of Farm Economy in Himachal Pradesh pointed out that diversified farming with vegetable production helped even the small farmers to earn sufficient income to make their livelihood. Similarly, a large number of other studies (IJAE, 1987; IJAE, 1988) show that small farmers too adopt multi-diversified farming, involving allocation of area under seasonal fruits, vegetables, dairy etc. for maximising their household income and employment, in almost all regions of the country. Nevertheless, it is also true that small scale diversified farming by marginal and small farmers do not generate adequate income for their sustenance in most cases (Haque, 1992).

Learning From Experiences of Other Countries

In recent years, a number of Asian countries where farm sizes are relatively smaller than that of India, have experienced agricultural diversification. Particularly the experiences of countries like South Korea, Indonesia and Philippines are quite instructive.

South Korean Experience

The smallholder agriculture of South Korea has witnessed a rapid rate of both horizontal and vertical diversification. The percentage share of agriculture in GDP declined from 40 percent in 1965 to 8 percent in 1992 and the proportion of agricultural labour force to total labour force declined from 55 percent in 1965 to 23 percent in 1992. Agro-industrial output is rising at the annual rate of about 13 percent, while the annual growth rates of output of fruits, meat, milk and fish also have been phenomenal to support horizontal diversification.

The first major effort for horizontal agricultural diversification was made as a part of Saemaul Integrated Income Growth Project launched in 1977, having emphasis on growth of orchard, native cattle, dairy cattle, vegetables, medicinal plants etc. The components of Saemaul Integrated Growth Project. The Project was implemented in consultation with the local farmers, covering 36,000 farm families and 2100 villages and fitted well into the regional environment. It had an integrated support mechanism through co-operation or team work for joint purchase of raw materials, joint sales of farm output, adoption of technology and economic use of farm by-products. In 1984, nearly 92 percent of farmers in the project area expressed the view that the project was helpful in raising their income through diversified agricultural growth.

In 1983, another Diversification Model Project was launched after reviewing the adaptability of the chosen crop, infrastructure growth, the degree of resource use, farm technology etc. The model project was initially selected by the country Planning Committee and agricultural co-operatives established project plan, provided funds and material assistance and sold farm products. While local governments provided public administrative services, rural guidance offices provided technological assistance.

As a result of the above mentioned efforts and their demonstration effects, the South Korean agriculture has been transformed from subsistence farming to commercial farming with rice, livestock, orchard and other cash crops. It has changed from multi-diversified farming with 9 to 10 crops to diversified farming with 3 to 5 crops which are more income generating and suit the regional environment.

In the past, small farms with multi-crop system were unable to control production and market price. But diversification was implemented at regional level with three to five specialised crops for each region which enabled them to influence regional production and prices and increase household income. The Diversification Model Project had the characteristics of ensuring (i) integrated growth of production, management and marketing, (ii) implementation of diversification plan at individual farm and regional level simultaneously and (iii) joint participation of government, agricultural co-operatives and farmers.

Indonesian Experience

The Indonesian economy is undergoing a process of rapid structural transformation in the sense that the share of agricultural sector in GDP declined drastically from 45 percent in 1970 to 13 percent in 1992. The share of agricultural employment declined from about 55 percent in 1985 to 47 percent in 1992. The average farm size is as low as 0.99 hectare and farm families possessing more than 2 hectares of land are very rare. The small farms in Java earn as much as 70 percent of family income from non-farm activities.

There is also sign of horizontal diversification. During the decade of 1980's, while the cereal production grew at the rate of 4.1 percent per annum, the annual growth rates of production of potato were as high as 10.4 percent and that of vegetables 6.5 percent, meat 8.1 percent, milk 8.3 percent and fish 4.8 percent. During the Fifth Five Year Plan covering the period from 1989-94, there was renewed emphasis on diversification of the rural economy, involving activities ranging from introduction of new technologies into existing farming system to growth of off-farm employment opportunities.

However, during the process of diversification, a number of constraints were faced, including (i) risk of crop failure due to insect-pest attacks, (ii) inappropriate plant protection techniques, (iii) inadequate access to credit for the cash-starved farmers, and (iv) inadequate access to quality seeds of high yielding varieties. The government as well as private sector were required to take necessary corrective measures. Nevertheless, the agricultural diversification programme in Indonesia can be said to have succeeded to a great extent in terms of growth of vegetables, meat, milk and fish production. In fact, future growth of Indonesian agriculture tends to lie in fisheries and livestock sub-sectors.

Moreover, given the present trends to continue, Java may become the centre of industrialisation and role of food production has to be assigned to outer islands which is not an easy task. Unlike South Korea, where local government, co-operatives and farmers played an effective role jointly in the process of rural diversification, in Indonesia, it was the Government which played a major role in diversifying agriculture by providing basic support facilities such as road, irrigation, marketing and preparation of commodity map for each region in order to utilise the available agricultural land in an optimal manner.

The Philippine Experience

The rural economy of Philippine witnessed mainly vertical diversification in recent years, having only a modest growth of horizontal diversification within the agricultural sector. The share of agriculture in GDP declined from 30 percent in 1970 to 22 percent in 1992 and the share of agricultural labour force decreased from 54 percent in 1970 to 46 percent in 1992. In terms of horizontal diversification, the share of crops in gross value added in agriculture marginally declined from 56.6 percent in 1980 to 53.2 percent in 1991. While the shares of rice and corn slightly improved, those of coconut, sugarcane, bananas and other crops declined. The share of livestock increased from 6.8 percent in 1980 to 10.5 percent in 1991 and those of poultry and fisheries increased from 4.8 percent and 15.0 percent in 1980 to 7.7 percent and 19.7 percent respectively in 1991. The contribution of forestry declined from 12.8 percent in 1980 to 4.1 percent in 1991.

The average size of holding in Philippines is as low as 1.4 hectares. Sugar and coconut oil are the two major agricultural export commodities. But in recent years, there has been a sharp decline in the export of these crops due to fierce international competition, particularly due to technological advances in sugar production in the west and the successful campaign carried out by the American Soyabean Association against tropical oils and large subsidies that soyabean oil exporting countries provide to their producers. In 1986, the Government dismantled all trading oligopolies, removed export taxes on agricultural commodities and lifted ban on the export of copra. But the export situation has not improved much. However, the adoption of more advanced technology and the favourable demand for aquaculture products in the international market and liberalisation of import of feed and feed ingredients for poultry helped in significant growth of these two agricultural sub-sectors.

Generally speaking, the agricultural diversification programme has been influenced by the following factors: It is almost mandatory that during wet season, rice will be grown in all irrigated areas, as the fanners want to be assured of meeting their rice requirements; A large proportion of wet land in irrigated areas may not be suitable for non-rice crops, unless there is land improvement or a significant technological breakthrough; The availability of credit facilities for crop production partly determines the kind of crop to be planted and area allocated for the crop; Availability of labour has not been a major constraint to diversification; The appropriate post-harvest technology for non-rice, perishable crops, is almost lacking which constrain the growth of non-rice crops; The Government has restricted import of certain commodities to protect domestic producers from unfair competition from agricultural commodities subsidised by exporting countries. These include rice, corn, sugar, beef, tobacco, onion, garlic, potatoes, cabbage and also coffee except in case of shortages; There is a favourable government policy which provides small farmers and fishermen greater access to credit and market for agricultural diversification; The agricultural diversification has so far emphasised more on food self-sufficiency and less on nutritional adequacy through fruits, vegetables and animal fats; The dairy growth programme as well as the production of sugar, corn and soyabeans in Philippines have suffered a set back due to unfair competition from developed countries which subsidise the production of these commodities. Can India Learn Any Lessons From the Experiences of Above mentioned Countries ?

India like South' Korea, Philippines and Indonesia can be broadly placed in the same category of smallholder agriculture, although the agrarian structure in India is a little more unequal than that in Korea and Indonesia. But India clearly lags behind the other countries in terms of institutional arrangements and effective delivery services at the grass root level.

In recent years, an effort has been made by the Planning Commission of India to prepare crop as well as overall diversification plan for various agro-climatic regions of the country, depending on the technological feasibility and economic viability of each enterprise. This may have to be further crystallised in consultation with farmers an scientists of concerned agro-climatic zones. India has also the benefit of a good net-work of agricultural research and extension organisation and a large number of Krishi Vigyan Kendras spread all over this country which can guide the farmers in technological matters. But India does not have an effective co-operative organisation at the grass root level to implement any agricultural diversification programme on a large scale.

Another lesson which can be learnt from the Korean experience is that small farm diversification should be broadly within the frame o1 3 to 5 selected, albeit relatively more productive and eco-friendly crops. Cropping patterns would vary according to various agro-climatic zones, thereby ensuring both the emergence of diversified farming system at regional and national levels and resource use efficiency at the farm level. However, the question remains whether we can organise our small farms along co-operative line for integrated growth of production, management and marketing and can rely on joint, albeit effective participation of farmers, panchayat and existing co-operatives in the implementation of diversification plan at individual farm and regional level simultaneously.

Indonesian experience however, was a little different. In Indonesia, it was the government which played a major role in diversifying agriculture through provision of basic support facilities and preparation of commodity map for each region in order to utilise the limited agricultural land in an optimal manner. In India too, the local self-government/panchayat at the village level are being strengthened and given a new growthal role. Therefore, on experimental basis, some diversification plan may have to be tried with the help of panchayat, local level administrator and Krishi Vigyan Kendras for further transmission of the model for application to other areas. The Philippine experience of diversification has been an admixture of both success and failure. It has strong farmers' association at the grass root level. But infrastructural facilities and post harvest technology for supporting non-rice crop led diversification are inadequate. Besides, like Indian farmers, the Philippine farmers too tend to stick to cultivation of major cereals for food security. Of course, India has something to learn from the failure of Philippines to promote crops like sugarcane and coconut for export, due to international market manipulations by developed countries, as it may happen to any country which relies mainly on export-led agricultural diversification.

Moreover, one common lesson from the experiences of Korea, Indonesia, Philippines and even Thailand is that small farms depend more on vertical diversification, involving growth of non-farm income opportunities than horizontal agricultural diversification.

EMERGING CONSEQUENCES, PROBLEMS AND PROSPECTS

In India, almost all small farms practise multi-diversified farming-and grow a number of crops even on small acreage and fragmented plots. Many of them undertake crop and livestock farming in a symbiotic manner. But this kind of farming does not necessarily yield enough returns for the sustenance and upward movement of small farm families. Therefore what is important for real diversification is that small farmers do not just grow a large number of less productive, subsistence crops and undertake traditional livestock farming. Instead, they should cultivate a few selected high yielding, high income generating and eco-friendly crops and livestock enterprises which fit well into the agro-climatic environment of each region. The data generated by the Agro-Climatic Regional Planning Unit of the Planning Commission reveal that there is immense possibility of diversified agricultural growth on small farms for which agricultural scientists, extension workers, administrators, planners, farmers' representatives and NGO's may have to work together to develop optimal farm and non-farm plans at the grass root level and synchronise them with regional and national level plans. Nevertheless, while planning for small farm diversification, we confront a number of issues and problems:

Technological Constraints

Technological change holds the key to agricultural diversification. The present level of technological growth for some of the high value fruits and vegetable crops is not only inadequate, but the available technology does not suit various agro-climatic situations. Moreover, the export led diversification would require technology for improving the quality of products, keeping in view the demands in domestic and international markets. There is need for growth of not only appropriate farm production technology, but also processing and marketing technologies for all those crops which are being considered for future diversification. A number of micro level studies have pointed out that non-availability of good planting material and high incidence of pests and insects act as major constraints to the adoption of some high value crops. Besides, variability in the yield of some of the high value crops are very high which would act as a deterrent to diversification. Therefore, generation of appropriate technologies including low cost biotechnologies and their quick dissemination would be essential for diversified growth of smallholder agriculture. The recent results of ICAR national demonstrations indicate that there is still a huge technological gap/yield gap in respect of important cereal crops in various regions of the" country. Bridging of such yield gaps for cereals crops may improve the prospect of non-cereal led diversification. India can even plan for competing with Thailand, Vietnam and USA for exporting surplus rice output to other countries, particularly if bridging of the existing yield gap reduces the cost of production per unit of output and makes the products competitive in the international markets. Agricultural research fund allocations in India are generally biased towards food and limited number of traditional export crops. The challenge of diversification demands far more concentrated research efforts for technology generation through scientific excellence and greater allocation of research funds in favour of high value crop and non-crop enterprises including fruits, vegetables, flowers, livestock, fisheries, agro-forestry, agro-processing, agri-business etc. Besides, substantial investment may be required in research on agricultural policy formulation and human resource growth, involving the training of scientists, extension workers and farmer leaders on the need and planned methods for agricultural diversification in various agro-climatic regions. Technological potentials are reported to exist. But bridging of the technological gap through further research, training and extension efforts would be crucial for diversified growth of small holder agriculture. This is intended to improve the productivity and quality of high value crops and their competitiveness in the domestic and global markets.

SHORT INSTITUTIONAL ARRANGEMENTS FOR DIVERSIFICATION

In a country like South Korea and Japan, the goal of diversification was successfully achieved primarily because of effective institutional arrangements at the grassroot level. A well co-ordinated effort of agricultural co-operatives, local self-government and rural guidance offices made the diversification programme in Korea a viable proposition. Can we think of some such institutional arrangements at the grassroot level which will provide the necessary credit, capital and marketing facilities? Can we revitalise our co-operative organisations for the purpose? Can the emerging Pancayat Raj system act as a catalyst? Can our Krishi Vigyan Kendras or extension workers be of some use? Alternatively, can private market entrepreneurship or a system of contract farming help promote diversification? What is our experience with the operationalisation and functioning of agri-business consortium? These are some of the issues which must be discussed at great length to work out a viable institutional arrangement to cater to the needs of small farm diversification.

Need for Vertical Diversification

One common feature of small farm diversification in countries like Japan, Korea, Indonesia, Philippines and even Thailand is that diversification plan of small farms invariably includes adequate growth of non-farm activities. In India also agricultural sector can hardly afford to sustain all its growing population and therefore, vertical diversification of small farms is all the more important. Small farms have to be basically part time farmers. But the investment and organisational requirements of such vertical diversification in the form of agro-industry, agri-business, agro-processing and services would be even greater. Also the role of Government, Private sector, Co-operative and NGOs in this regard need to be properly defined.

Physical Constraints

All types of lands and locations are not regularly suitable for diversified, albeit profitable alternative farming. Moreover, in greater part of high rainfall and irrigated zones, heavy textured soil and poor drainage system stand in the way of diversification in favour of non-rice crops, unless land improvement and drainage measures are undertaken.

The question is whether efforts of state government, panchayat or collective action of the farmers would be forthcoming for such land improvement? Even in the dry areas, diversified farming would depend much on soil type, topography etc. Also fragmented plots of small holdings act as a constraint to efficient use of land for diversification, particularly because the farms get deprived of the necessary economies of scale. Therefore, consolidation of even smaller holdings along with provisions of joint farming could help promote diversification.

In the past, some states have not implemented consolidation of holdings on the pretext that this leads to eviction of tenants and small and marginal farmers. But this problem could be solved by protecting the land rights of each category of farmers.

Need for Policy Support

The traditional agricultural growth policies in India have not been favourable to diversified agricultural growth, particularly in the context of small farms and backward regions. Therefore, if our objective is to promote small farm diversification in various agro-climatic regions of the country, the existing policies may have to be partly modified. First, the agricultural price policy does not provide adequate protection in terms of procurement and minimum support prices to fruits and vegetable crops. Since the production and prices of these commodities fluctuate widely, small farmers would find it difficult to grow these crops, unless price protection is available.

Secondly, small farmers, particularly in backward regions do not have access to cold storage, market and transportation facilities. Similarly, in the absence of stable power supply, the promotion of non-farm activities like agro-processing would be unthinkable. Primarily, it should be the responsibility of the Government to develop these infrastructural facilities, although willing private entrepreneurs could also be involved.

Thirdly, tenancy laws of most of the states may have to be modified to allow the large farmers to take up non-farm activities and to enable the marginal and small farmers to have greater access to land through lease market. This would enhance their economic viability and also promote diversified agricultural growth. 'Similarly, consolidation of holdings of even smaller holdings would enable them to enjoy better economies of scale.

Fourthly, technology generation policy would need greater support in terms of investment in research, extension and training in favour of those crops and enterprises which are being considered for diversification.

Fifthly, in view of high variability of production, prices and income, the possibility of crop insurance coverage for some agricultural commodities which are perishable but of great social value due to their nutritional potentials and foreign exchange earning capacity may be considered.

Finally, the future direction of small farm diversification in India would largely depend on the direction and effectiveness of government policy that could influence the creation of appropriate socio-economic environment for diversified farming through necessary technological, infrastructural, institutional and administrative changes.

The latest available data show that still about 66 per cent of the country's labour force is engaged in agriculture and share only 32 percent of GDP. Conversely, 34 percent of the labour force being in the non-agricultural sector share about 68 percent of GDP.

Such glaring income disparities between agricultural and non-agricultural households must be removed through appropriate policy changes for horizontal as well as vertical diversification in rural areas. Indeed, the prospects of small farm diversification in India are quite high. What is needed is a real commitment on the part of our policy makers, administrators, scientists and extension workers to make small farm diversification a feasible and viable proposition.

BACKWARD AGRICULTURE-A CAUSE OF INDIA'S POVERTY

The biggest challenge facing India's policy makers is the persisting high incidence of poverty. One of the reasons for the high incidence of poverty in India is its backward agriculture, whose productivity per hectare and per capita is amongst the lowest. The low per capita production is not only due to huge population, but also due to low productivity, which is only 64 percent of the world average. The productivity potential of Indian agriculture has been amply demonstrated in Punjab, where it is nearly equal to those of developed countries. If India attains the productivity level already achieved in Punjab, the national income of India would be more than double of what it is today. Taking into thought the vast untapped potential of Indian agriculture, the pool of unutilized scientific knowledge and the eagerness of Indian farmers respond to opportunities, neither Indian agriculture should remain backward, nor our people should remain almost the poorest in the world. Quite often, low productivity of Indian agriculture is attributed to small size of landholdings. In Japan, Korea and Taiwan the average size of holding is nearly the same as in India, but their incomes per hectare and per worker are several times more than ours.

The Decisive Factor

Among the many factors which determine productivity in agriculture, the most decisive one is the kind of "social support" which agriculturists get from the rest of the society. Social support emanates from a sense of gratitude towards agriculturists who provide the most essential requirement of mankind. This, however, is not quantifiable are therefore not comparable amongst various countries of the world. Hence a new concept, which takes into account only monetary transfers has bee developed. This is called, "Producer's Subsidy Equivalent" PSE. The can be defined, as payments received by agriculturists, in excess of what they would have received, if the Government had not intervened in free marketing of agricultural produce. PSE includes direct and indirect payments by Government and also increased payments to farmers b consumers, made inevitable by imposing ban on import of cheaper farm products from abroad.

The U.S. Department of Agriculture in their publication entitled GATT & Agriculture" has reported the average PSE for 17 countries, for a period of 5 years from 1982-86. The highest PSE (plus 70 percent) was in Japan, followed by S. Korea,Mexico, E.E.C and Canada. In U.S.A. it was above 25 percent. It means that in these countries, farmers received 25 to 70 percent more than what they would have received under conditions of free trade. In four countries, viz Pakistan, Nigeria, India and Argentina, PSEs were negative. In India it was below 18 percent. It means farmers in Indiareceived 18 percent less than what they would have received, if the Government had not intervened in free trade. This has been arrived at after deducting from the total losses suffered by farmers due to Government interventions and considering the amounts spent by Government for providing to farmers input assistance and infrastructural support. It is significant that in all the countries, in which PSEs were found to be negative, productivity of agriculture was much below the average of the world. In other words, wherever farmers have been exploited, agriculture has remained backward. It is not necessary to plead for subsidy to farmers in India, comparable to what their counterparts are getting in the developed countries. But anti-farmer interventions by the Government in free marketing of farm products, must now be stopped. Unless this is done, profitability, savings and capital formation in the farm sector will remain low and potentials of Indian agriculture unutilised.

The following quotation from Economic Survey, 1994-95 is quite instructive. "Gross investment in real terms (at 1980-81 prices) in agriculture has stagnated. It was Rs. 4,636 crore in 1980-81 and Rs. 4,617 crore in 199^2-93 (actual). The total gross domestic formation declined from 18 percent in 1980-81 to 9 percent in 1992-93. The decline in real capital formation in agriculture in the public sector is more perceptible, as it has come down to Rs. 1,065 crore in 1992-93 compared to Rs. 1,796 crore in 1980-81. Private sector real investment in agriculture has increased in absolute terms from Rs 2,840 crore in 1980-81 to Rs.3,552 crore in 1992-93, though its share in total gross capital formation has declined significantly during the period. This decreasing share of private investment in the total gross capital formation seems to suggest that the agricultural sector is relatively less attractive for private investment as compared to other sectors of the economy. This suggests a need to strengthen incentives for attracting more private investment into agriculture."

Small Holdings

The average size of small holdings, measuring less than two hectares, is 0.6 hectare and the annual income required to support a family above the poverty level is now estimated to be Rs. 12,000. Therefore, the crucial question is: Can a land-holding of 0.6 hectare generate an annual income of more than Rs.12,000 ? The reply to this question is both "No" and "Yes". Income from farming does not depend only on the size of the farm, but on the capital invested on it, both in monetary terms and in technological know-how. Many small farmers in developed countries are earning many times more than what most land holders of 18 acres are earning in India. No doubt, glass-house is an extreme example. But there certainly is the possibility of earning an annual income of more than Rs. 12,000 from 0.6 hectares of land, provided the farmer gives up conventional farming i.e., production of grains and takes to horticulture and animal husbandry. Fortunately, the Government does not intervene in marketing of fruits, vegetables and animal products. Hence their prices have risen to about the same level, as the prices of all commodities. While the terms of trade are against producers of grains, that is not so, in case of producers of fruits, vegetables and animal products. One can surely earn net profit of more than Rs.12,000/-annually from a holding of 0.6 hectares by producing fruits, like banana, papaya, guava, lemon, mango, litchi, grapes, and vegetables like hybrid brinjal, tomato and cabbage. Plant propagation and seed production are even more rewarding. Animal husbandry, poultry keeping, goat keeping, pisciculture and dairying are no less profitable. These enterprises can also be combined. But for success in these specialised lines of agriculture, lot more capital, expertise and infrastructural facilities are required. To enable small farmers to take specialized farming, certain systemic changes need to be made. The are briefly stated below:

Institutional Credit

Institutional credit at present is advanced either on the basis of seasonal needs of conventional farming or of the value of land owned by a farmer. In case of small farmer, the value of land is naturally less, but his credit needs will be more if he takes to animal husbandry, poultry or fishery. Similarly, his repayment period will have to be longer, if he takes to production of fruits. Therefore, institutional credit will have to tailored according to the needs of small farmers.

Area Specific Programme

There is no lack of techniques of production, even for dryland and small farms, which can generate more incomes and employment in villages. But these cannot be adopted by small farmers, unless the necessary services to support these enterprises in the form of backward and forward linkages are first provided. Rural India will remain poor so long as (1) new techniques of farm production are not adopted more widely, (2) adequate capital in the form of loans for adoption of new techniques is not provided and (3) efficient marketing arrangements for the produce are not made which reduce the cost of distribution. All these will have to be provided simultaneously in a package and not in a haphazard manner.

As it will not be possible to provide all the different services needed to support the different enterprises, it is advisable that farmers in every growth block of the country be asked to select one cash crop or enterprise most suited to their area. After this decision has been made, every assistance-technological, financial and managerial-should be provided for that particular crop or enterprise through co-operative or panchayati effort. The NDDB has done a commendable job in providing all the facilities to milk producers of Gujarat. Similar facilities need to be provided for promotion and production of horticultural and animal husbandry products and in course of time, also for promotion of agro-processing industries and export of such products.

Model Farmer Scheme

To promote the above area-specific programme, it is necessary to promote participatory extension strategy. For this, it is suggested that at least one model small farmer be selected in each village with a population of more than 500. The model farmer should be educated upto high school and own land less than 2 hectares. His selection should be made strictly on merit. After selection, he should be trained at Government cost at a research institute for a period of time, considered necessary to train him in the use of the latest techniques of production for that crop or enterprise, which has been selected for his particular area. But mere training will not suffice, as it has been repeatedly proved that knowledge without capital is not of much help. Therefore, the model farmer should also be equipped, at Government cost, with whatever is necessary to enable him to apply the techniques of production, which he has learnt. The amount spent on equipping the model farmer with improved animal and tools of production should berealised from him in 25 yearly equal instalments, calculated at a concessional rate of interest. These model farmers, if properly trained, are likely to prove far more effective in dissemination of the new methods of intensive farming than the village level workers. Farmers learn more by seeing what their neighbours are doing than by listening to lectures. These model farmers will serve as pioneers of intensive farming as well as links between researchers and users of research. This scheme will enable about 3 lakhrural youth to earn a decent living, not at Government cost but through training and their hard work. Whatever the Government spends initially on their training and equipment, will have to be realized from them later on.

The model farmers, if trained and enabled to produce improved seeds and plants, will also effectively meet the shortages of these inputs in the country which Government agencies have so far failed to do. The model small farmers can also be trained in artificial insemination of cows and plant protection to render these services to their neighbours.

Reorientation of Agricultural Research

Agricultural research institutions should now focus their research efforts on increasing the overall productivity of marginal and small farmers. They should also pay more attention to dryland as well as low land farming. So far, efforts have been mainly directed to obtain maximum production under the most favourable conditions. It is one thing to increase productivity on large farms through optimum use of modern inputs and quite another to make small farms profitable under difficult conditions. In India, 80 percent of the cultivators operate less than two hectares of land and majority of them also do not have adequate irrigation facilities. It is towards them that the researchers should now direct their attention. Plant breeders should aim to evolve varieties which are better suited to face environmental stresses such as drought, flood, salinity and diseases. Also more attention should be given to bio-fertilization of soil and regenerative agriculture.

7

Agricultural Development and Rural Poverty

INTRODUCTION

The provision of public goods and the related issue of ameliorating the impact of market failures are the classical justifications for government intervention in the economy. In reality governments do not do a good job in fulfilling these roles. (1) Governments intervene in markets that cannot be even remotely considered affected by failure or imperfections. (2) They spend a large share of the financial, human and institutional public resources on private goods (subsidies), not on public goods. Misguided public policies that interfere with essentially efficient markets have been for a long time a central concern to economists.

The literature documenting and measuring the efficiency costs of market interferences for many commodities and many countries around the world is indeed massive· On the basis of these studies international organizations as well as several other economic institutions have forcefully argued for the removal of government interference with markets operation for several decades.

The high degree of attention that the economics literature has given to issue (1), contrasts with the relatively little attention that issue (2) has received from both academic and policy economists. This despite the potentially large efficiency and equity losses arising from misguided allocations of public expenditures. This chapter empirically documents the size of (2) in the rural sector and investigates its consequences for rural development (agricultural growth, patterns of growth and rural poverty) for ten Latin American countries over 1985-2000.

The fact that raising government revenues is costly and that the availability of government financial, human and institutional resources is usually tight, implies that deviating part of these resources to private goods crowds out the supply of public goods. Spending large volumes of public resources in private goods means that governments are over taxing the private sector, or under supplying public goods or both. Whatever the means used by the public sector to finance spending in private goods, it implies efficiency losses associated with the inevitable distortions caused by raising more taxes than needed or by under providing public goods.

The under provision of public goods may have serious consequences for the productivity of private investments as public goods are important complementary assets with private capital. A low supply of public goods means scarcity of human capital, under investment in research and development (R&D), less infrastructure, insufficient environmental protection, etc. (World Bank, 2000). All these are important assets that contribute to increase the productivity of private investments. Hence their scarcity is likely to constitute an obstacle to economic expansion.

The biased structure of public expenditure allocation may have important equity implications as well. For reasons to be discussed below, government expenditures in (non-social) subsidies and other private goods tend to be directed to the wealthier segments of society to the detriment of the poor. Over taxation that spending in private goods may cause, also has negative equity consequences. It is well known that in most developing countries taxes are not progressive, especially because of its great reliance on indirect taxation. So the additional taxes needed to finance subsidies to the wealthy are in a significant part paid by the poor. In addition, the under supply of public goods affect the welfare of the poor much more than the welfare of the wealthy. The only significant assets of the poor are their labour force and, for the rural poor, their natural resources. The income of the poor is thus extremely dependent on their ability to enhance their human capital and to protect the natural assets, both important public or semipublic goods. There have been many empirical studies that have measured the impact of specific public goods such as R&D, rural roads and other infrastructure on agriculture as well as on other sectors. What seems to be missing, however, is a broader approach that may permit evaluating the impact of public goods vis-à-vis other types of government expenditures on efficiency and social equity. Without such a comparison the conceptual and policy implications of findings pointing to the high pay-offs of certain specific public expenditures is in part obscured. At the policy level, there seems to be a degree of consensus regarding the importance of an adequate provision of public goods for economic expansion and social equity. However, the same lack of a clear conceptual framework and the general weakness of empirical studies due in part to such a conceptual deficit also mean a certain weakness in the policy recommendations. It is clearly a rather empty policy statement to advise governments to increase expenditures in education, health, roads, etc. without pinpointing the source of the low expenditures in such goods in the first place and without indicating what expenditures to cut (or what taxes to raise) in the second place. In any case, the concerns shown by the national and international institutions about issue (2) appears to be puny compared to the overwhelming focus in reducing government market interventions. Below we provide evidence suggesting that such lack of balance is not justified.

Apart from affecting the rate of agricultural growth, the heavy emphasis of governments on subsidizing the wealthy (and thus under supplying public goods) is likely to affect the patterns of agricultural growth. In particular, a hypothesis that we test here is that (non-social) subsidies tend to make agricultural growth much more dependent on land expansion than on intensification.

In many countries, especially in tropical ones, agricultural growth is much more linked to area expansion than to intensification. This pattern of growth is particularly deleterious for forests and is often responsible for the large deforestation impacts attributed to agriculture by many studies of deforestation (López and Galinato, 2004). When governments are more prone to provide non-social subsidies it has been shown that expanding farm size beyond optimal levels is a rational response by farmers as a way of signalling their disposition to offer bribes in exchange for subsidies (Bulte et.al., 2004). Land expansion is thus a vehicle to promote subsidies and knowing that governments are prone to subsidize induce farmers to expand land area. That is, land expansion and subsidies are cause and effect of each other. Without claiming to derive causal links, we empirically test for the existence of a positive correlation between subsidies and land expansion.

The objective of this chapter is to shed light on the following questions: What is the effect of increasing rural public expenditures on agricultural income and rural poverty when such expenditures contain a high share of expenditures on private goods? What would be the impact of changing the structure of rural public expenditures (e.g., reducing expenditures on private goods and concomitantly increasing public good expenditures while keeping total expenditures constant) on agricultural income and poverty? To what extent subsidies promote greater reliance of agriculture on area expansion? How does the impact of such policies compare with the effect of traditional policy prescriptions associated with reducing certain market interventions?

A brief conceptual framework and hypotheses are first presented. Next we review existing empirical evidence suggesting that the issues raised are likely to be of great policy relevance. Then we present a quantitative description of the evolution of non-agriculture factors likely to affect the performance of agriculture over the period 1985-2000 for ten Latin American countries. The countries considered to have consistent rural public expenditure series with a degree of detail sufficient to allow us to aggregate them into expenditures in public goods and in private goods. We show that for the countries considered to have a large share of total public expenditures in the rural sector, it is mostly spent on private goods. Next we discuss the econometric model, and in the following section the econometric results are presented. We conclude with an evaluation of the empirical findings as a contribution to the development of a policy agenda that is broader than the conventional one.

TARGETING MEASURES USED IN ANTI-POVERTY PROGRAMS

There exist a large number of small and big PTPs in India, channelled through different ministries of the central government, and with different modalities of implementation. Some of these schemes are implemented by the state governments while others have larger proportion of funds flowing directly from the centre to district administrations. Obviously, a comprehensive review of each and every PTP would be neither feasible nor desirable. The discussion below therefore centres on a select subset of the PTPs, based upon their relative quantitative importance, availability of information relating to their implementation monitoring and evaluation, and relevance to the objectives of the present analysis, based upon targeting design and effectiveness of the scheme. The schemes chosen include examples of pure income transfer, food for work, self-employment and infrastructure generation.

First, however, a brief discussion of targeting mechanism in the Indian context is useful. A widely used categorization of targeting mechanisms that can be used to classify PTPs is as listed below.

- Self-targeting:: Such schemes rely on differential incentives of agents in tackling the problem of asymmetric information between the principal (the government providing poverty relief) and the agents (households or individuals affected by the government schemes). The design of the schemes has the objective of making the scheme worth participating in only for those who are poor, not for others.
- Activity targeting: Relies on "broad" targeting, primarily through subsidized provision of goods and services whose benefit incidence will be progressive, i.e., falling largely on those poor rather than better-

off. Examples typically include primary education, provision of primary health care and basic health services in rural areas, and broadly targeted subsidies for irrigation, power and fertilizers.

- Location targeting: Based on geographical distribution of poverty, seeking to target interventions in geographic areas with high concentration of the poor.
- Indicator targeting: Relies on non-income indicators that are meant to be correlated with poverty. These can include lack of or size of ownership of land, form of dwelling, social status, gender of head of household, etc.

Asymmetric information between the government, seeking to provide transfers to the poor, and individuals or households in the economy who can legitimately or otherwise seek these transfers, is the *raison de étre* of targeting. The underlying rationale of these targeting mechanisms is that administrative and other costs of identifying those who are poor are high, potentially reducing the resources that would be transferred to the poor under the scheme. Targeting mechanisms are a contractual/program-design innovation in response to the information asymmetry and the high costs of overcoming the information barrier.

However, this framework is implicitly less than comprehensive in approach, in the sense of focusing only on one scheme at a time. In a context where the principal (in a principal-agent context) has several schemes in operation, the administrative costs *per scheme* (of overcoming information asymmetry) can get diluted substantially, thereby vitiating the need for indirect targeting mechanisms for any specific scheme.

Put alternatively, the issue of whether or not the administrative costs of identifying the poor are undertaken by the government usually does not depend on any specific scheme. In an inter-temporal context, where the government does not know what specific schemes it may want to implement in near future, "tagging the poor" – Administrative Identification – may provide externalities in terms of greater choices of schemes and their designs. This is an important issue, as shown by the Indian experience where a large number of government poverty-targeted schemes rely on "Administrative Identification" (AI) to select beneficiaries. The most common criterion used in government schemes is that beneficiaries should be households below the poverty line (or BPL households).

Other criteria, such as focusing on SCs/STs (which per se would represent indicator targeting in the Indian context) are overlaid on the BPL status. As mentioned above, it may be argued that with an aggregate annual budget on

CSS schemes exceeding Rs. 250 billion, it may be worthwhile for the government to undertake AI to better target the poor. Indeed, analytically it is perhaps more pertinent to ask why other targeting mechanisms should exist at all once AI has been undertaken. Use of indirect targeting mechanisms in conjunction with AI may reflect in part the recognition that implementation of AI may be imperfect due to various reasons. In particular, the process itself may suffer from high Type I and Type II errors, as discussed below, resulting in exclusion of many poor and inclusion of many non-poor. In addition, the frequency of identification is necessarily spread apart in time, which would make it impossible to differentiate between transient and chronic poverty, (e.g., to differentiate the needy seeking food for work in face of natural calamity).

Administrative Identification: Tagging BPL (Below Poverty Line) Families

Since most PTPs currently in existence directly or indirectly rely on administrative classification of households into BPL and APL (Above Poverty Line), it is useful to briefly explain how this identification is undertaken. The exercise is intimately related to government efforts to provide food security to the population through the Public Distribution System (PDS). The PDS, in its earlier forms, dates back to almost fifty years ago and was a general entitlement scheme with universal coverage until 1992. It provided rationed quantity of basic food (rice, wheat, sugar, edible oils) and some essential non-food items (kerosene oil and coal) at prices substantially below market prices. The central government was responsible for procuring, storing and transporting the PDS commodities up to central warehouses in each state/union territory, while the state government was responsible for distribution within the state.

While the universal coverage of PDS continued, the government introduced two major changes, the first in 1992, in the form of the Revamped PDS (RPDS) and, subsequently, in 1997 as the Targeted PDS (TPDS), both innovations targeted at poor households. The RPDS relied on geographical targeting, being introduced with universal coverage in only 1775 blocks in poor areas – mainly tribal and hilly, drought prone and remotely located areas. The TPDS, on the other hand, was implemented in all areas but was open only to those identified as BPL. Along with the introduction of the TPDS, the price differential between PDS shops and open market was almost eliminated, effectively providing subsidized food only to BPL families.

At the core of the TPDS was division of the entire population into BPL and APL categories, based on the poverty line defined by the Planning Commission of India for different states for 1993-94. Multiple criteria were adopted for classification of BPL households, which in addition to income also included qualitative parameters like household occupation, housing conditions, number of earners, land operated or owned, live-stock, and ownership of durables such as TV, refrigerator, motor cycle/scooter, three wheelers, tractors, power tillers, combined threshers, etc. The responsibility for undertaking surveys and identifying the poor was with the state governments. However, the total number of BPL families in each state was capped somewhat arbitrarily at state-level estimates of the poor made by the Planning Commission using data for 1993-94, adjusted for growth in population in the interim.

RURAL EMPLOYMENT PROGRAM, INCLUDING FOOD FOR WORK - SGRY

The SGRY, targeting poverty reduction through employment generation, has a long history in India, spanning several incarnations. Its genesis lies in the National Rural Employment Program and the Rural Landless Employment Program, both of which were initiated in the early 1970s, but subsequently merged into a new scheme called Jawahar Rogzar Yojana (JRY) or the Jawahar Employment Scheme in 1989. The JRY was meant to offer additional gainful employment to the unemployed and the under-employed people in rural areas through creation of rural economic infrastructure. However, the program fell short of objectives in several ways. Employment generation was too inadequate to be meaningful, with an average of roughly 11 days of employment created per person according to an evaluation in 1994 (GoI (2000)). The resources available were spread too thinly so as to increase coverage without concern for duration of employment. The asset creation involved high material costs and was not particularly labour intensive, in flagrant violation of prescribed norms. Other routine violations included the use of contractors or middlemen who often hired outside labourers to lower the wage rates, and used trucks and tractors instead of labour intensive techniques. Fudging of muster rolls and measurement books was thus rampant. Only 17 percent of jobs generated under the JRY went to women, against a target ratio of 30 percent.

According to estimates presented by Dev and Evenson (2003), the cost of transferring one rupee under the JRY was rupees 2.28. This can be compared to a cost of rupee 1.85 per rupee transferred under the Maharashtra Employment Guarantee Scheme, a relatively better run program and a high ratio of rupees 6.68 per rupee transferred under the PDS. In view of its deficiencies, the JRY was restructured and transformed in 1999 into a new scheme – the Jawahar Gram Samrudhi Yojana (JGSY) or the Jawahar Rural Advancement Scheme, which explicitly prioritized asset creation as the primary objective, followed by wage employment. No evaluation studies of JGSY are available since it lasted only a short time before being merged into a new scheme, the SGRY.

The transformation of the JRY into the JGSY was preceded by a parallel scheme launched in 1993 – the Employment Assurance Scheme (EAS) – that had similar objectives as the JRY/JGSY but with reversed priorities. The EAS, which was in implementation through most of the country by 1997-98, had the primary objective of creating additional wage employment for rural poor through manual work in periods of acute shortage of employment opportunities, with asset creation as a secondary objective. The EAS relied on self-targeting by setting of wage rates below market wages. However, the EAS showed similar failings in implementation as its close variant, the JGSY. For example, the scheme generated on average only about 17 days of employment per year per person according to a study by the Controller and Auditor General of India (GoI (2000)). The objective of the EAS was, by comparison, to provide assured employment of 100 days per year at statutory minimum wages. The self-selection targeting was subverted by routine use of contractors in most states, fudging of the employment rolls, and violation of norms that called for a 60:40 split of wages and materials in asset creation. As a result, in three states – West Bengal, Gujarat and Haryana – the estimated unit cost of generating a day's employment was Rs. 200 to Rs. 300, far in excess of wage rates. For the country as a whole, the mid-term evaluation estimated only Rs. 15 of every 100 rupees' expenditure reached the beneficiaries wages, against a target of Rs. 60. No inventory of assetswas kept, making it difficult to ascertain whether assets created were community assets or for individual benefit. In addition, with deteriorating finances of state governments, allocated funds did not reach the ground in many cases, in part due to lack of matching funds from state governments.

Amongst similar problems in the implementation of the EAS, a review by the Planning Commission (PEO(2000)) found effective rates of utilization of funds was only about 67 percent of the notional minimum allocations of administrative blocks. This reflected in part lack of matching funds from state government preventing release of funds, and also *ad hoc* and untimely release of funds by the governments. For example, the first part of central government allocation (40 percent of the total allocation) was to be provided at the beginning of the financial year, with the remaining to be released at receipt of utilization certificates. In practice, however, in more than half the 14 states studied, states received more than 50 percent of their allocation in the last quarter of the year. There was also evidence of significant diversion of funds, reflected in mismatch between allocation and expenditure of funds at different nodes of implementation of the scheme. This diversion was also noted by the report of the Controller and Auditor General of India (CAG (1997), No.3). PEO (2000) also found only 32 percent of villages were covered by the EAS in an average block, with little consistency in implementation within specific villages. Thus, only 5.4 percent of villages covered in a block typically had the EAS operations in each year during the first four years of the EAS (1993-97). Thus, coverage of villages within specific blocks was *ad hoc*, allowing discretion to district administrations, and there was little credibility in terms of providing assurance of employment in the villages actually covered. In addition, the study estimated that on average the EAS implementation covered only 16 percent of the target group in the chosen villages. Thus, the effective annual coverage of the target groups in ten of the 14 states was less than 10 percent, being as low as 1 to 3 percent in some states. Combined with the small number of days of employment generated per person on average, the impact of the EAS on household incomes was negligible. Although the EAS was quantitatively more significant as a source of income than other government wage employment programs running in parallel in the villages, it contributed only 11.5 percent of household annual income to the extremely limited group of selected beneficiaries.

Thus, in practice, there was little difference between the JGSY and the EAS, in terms of both objectives and implementation failures, with the only substantive difference being administrative. The JGSY was being implemented by the village-level institutions (PRIs) while the EAS relied on the state administrative apparatus. In September 2001, on-going schemes under the EAS and the JGSY were merged into a new scheme – the SGRY. The objectives of SGRY are to provide additional wage employment in rural areas as also food security, alongside the creation of durable community, social and economic assets and infrastructure development. A part of wages to the workers is to be distributed in the form of 5 kilograms of food grains per manday. The cash component is shared by the central and state governments in the ratio of 75:25, while the cost of food grains distributed to the states is borne entirely by the central government. The SGRY is implemented in two streams with each stream receiving half of the total resources available. The first stream is implemented through District and intermediate elected bodies (DRDAs/*Zilla Parishads and Block Panchayats*) while the other stream allocates funds to the village panchayats.

The SGRY also encompasses all "Food for Work" programs in the country, since it includes a special component for augmenting food security through additional wage employment in calamity affected rural areas. A certain percentage of foodgrains allotted under the SGRY is reserved for this purpose. Food grains under the Special Component can be utilized by any scheme of the central or state government that is implemented to generate additional employment in calamity affected areas. The cash component of the costs and costs of materials are, however, met from the scheme under which the Special Component is utilized.

Thus, administrative arrangements for implementing the SGRY involve coordination among three central ministries, namely, Agriculture, Food and Rural Development. The Food Ministry releases grains at the direction of the Ministry of Agriculture, while the Rural Development Ministry is responsible for administration and supervision. The scheme is self-targeting and available to rural poor (BPL or otherwise) who are in need of wage employment and willing to take up unskilled, manual work at specified wage rates. Preference is given to the poorest of the poor, women, SCs/STs, and parents of child labour withdrawn from hazardous occupations. The beneficiaries are selected by *Gram Panchayat* during meetings of the *Gram Sabha.*

The emphasis on payment in kind-via foodgrains-combined with the difficulty and cost of storing and transporting foodgrains reduces the scope of misappropriation of resources by officials. Nonetheless, responsibilities for storage, transport and distribution within districts are contracted out to "fair price" shops (that are part of the central government's targeted public distribution system (PDS) of subsidized foodgrains) and the associated contractors. This has created substantial scope for fraudulent practices due to large gap between market prices and prices in fair-price shops (though the gap has been declining of late). Nayak et.al. (2003) estimate that the combination of malpractice among administrators at lower levels of government and contractors results in perhaps only 25 percent of the wage funds to which beneficiaries are entitled actually reaching them.

SGRY, the latest incarnation of employment-based, food-for-work PTP in India is too recent in vintage for any comprehensive evaluation studies to have been done. However, one recent study has reviewed the implementation of the SGRY in the state of Andhra Pradesh. Using primary data collected over 12 months between 2001 and 2002 from 6 villages, Deshingkar and Johnson (2003) find little has changed at the ground level in implementing the SGRY. Despite attempts at decentralization of decision making in the SGRY, village-level government (*Gram Sabhas and Panchayats*) are often controlled by the local (landed) elite. Beneficiaries were selected during meetings among the local officials (members of *Panchayat, Sarpanches, Mandal* officials) and contractors, and the decisions announced in the village meetings of *Gram Sabha*. In three villages, the largest number of labourers were hired from the hamlet of the *Sarpanch* (village head), while in another the largest share of hired labour belonged only to the caste of the village head.

There was also widespread use of contractors, contrary to the scheme guidelines, often in connivance with local officials. The contractors also obtained illegal profit by claiming the full food grain (rice) quota for partially and poorly completed works, claiming rice for old assets already completed under some other scheme, 'double-dipping" by filing separate claims to different departments for same work, and submitting inflated costs in works proposals. In addition, contractors often decided to pay labour only in cash since rice was released late to the contractors, or the contractor could sell the rice in the open market at profit.

The impact of the scheme varied across the six villages, being relatively high in one village (which also had the lowest corruption in implementation), but miniscule in most others. On an average, 24.4 percent of households sampled in these villages had participated in the scheme (ranging from 65 percent in one village to 3 percent in the worst case). With the exception of the single village, the number of person days of employment created was also low, averaging lessthan 14 days per participating household. Two major reasons contributing to this, aside from corruption and leakage, were use of outside/migrant labour by contractors to minimize costs, and substantial use of labour-displacing machinery (often owned by contractors).

This brief review of employment generation and food-for-work type of PTP in India paints a fairly bleak picture, and several problems in implementation. Some of these problems are well known, such as the importance of appropriate wage setting in affecting screening efficiency. In case of SGRY in Andhra Pradesh, Deshingkar and Johnson (2003) document wages were too low in relatively prosperous villages, leading to use of migrant labour and machinery, while in poor villages the wages were much higher than prevailing rates, leading to crowding out of the really poor. A well designed scheme of self-selection (without quantitative rationing) should lead to virtually negligible levels of Type I and II errors, in that only the really poor would be willing to work at wages below prevailing rates, and all those willing to work at these rates would be accommodated. With only quantitative rationing (cap on total funds available), some needy poor may not be able to access wage employment leading to Type I error in targeting. In relatively prosperous villages, with wages set too low, local labour may not have participated but presumably the migrant poor willing to work at these wages were a justifiable target, implying little leakage in terms of Type I targeting error. In contrast, in poor villages where the PTP was really needed, the higher than prevailing rates created room for Type I errors, with the poor crowded out.

However, problems other than wage setting are of deeper and greater concern, stemming from institutional and governance constraints, and magnifying both types of targeting errors, II and I. While the government has sought to decentralize scheme implementation to create greater ownership of resulting assets, the local level administration and the PRIs are strongly susceptible to corruption. This has resulted in flagrant violations of government guidelines, including use of contractors/intermediaries, excessive reliance on labour-displacing machinery, payment in cash instead of kind, and doubtful selection/quality of the assets created. Greater reliance on labour-displacing machinery and payment in cash rather than cash and kind magnifies Type I errors leading to exclusion of benefits of those who are really poor at the expense of funds diverted to owners of the machinery and middlemen. Other corrupt practices magnify Type II errors by diverting funds and benefiting corrupt officials and middlemen who are not the intended beneficiaries of the program. Thus, even the most well-designed scheme, relying on the iron logic of self-selection will fail in implementation if employment rolls can be falsified, assets shown as created when they actually are not, and payments made that are below those legally mandated. Although employment generation using food for work continues to be a critical element of poverty targeting in areas adversely affected, the leakage of funds, corruption and poor governance result in the impact of these interventions on poverty being substantially diluted.

Employment Guarantee Scheme of MaharashtraAn early food-for-work program implemented in the western Indian state of Maharashtra was widely regarded as very successful in its initial years, and indeed was the inspiration for the EAS introduced at national level. The Employment Guarantee Scheme (EGS) started on a pilot basis in 1965 in one district (Sangli) and a modified EGS was implemented across the state in 1972, following one of the most severe droughts in the region in recent history. The scheme was soon suspended for two years, replaced by central government schemes, but in 1974 the state government decided to set up a permanent scheme using only state resources, leading to resumption of the EGS. It was provided a statutory basis with the enactment of the Maharashtra Employment Guarantee Act of 1977. The scheme is financed by urban taxes (on professionals and motor vehicles) with matching grants from the state government. The EGS is unique for several reasons, being one of the oldest such schemes in developing world, its large scale of operations at inception, and the fact that it guarantees employment (rather than merely assuring it). The EGS provides a guarantee of employment to all adults above 18 years of age who are willing to do unskilled manual labour on a piece-rate basis. Its primary objective is, thus, creation of employment opportunities with the secondary objective of creating rural assets to provide drought proofing, soil management and conservation.

Starting from 4.5 million person-days of jobs created in its first year, the EGS was generating more than 100 million person days of employment by early 1980s (190 million in 1986), before declining to 80-90 million person days after 1989 (Dev (1995)). Cumulatively, the scheme spent Rs. 27 billion up to 1991 to create about 2.3 billion person days of employment in the state. Not surprisingly, the EGS is one of the most analysed public-works programs in the literature, and received high marks in its initial years from most evaluations.

Although quantitative data on type II errors due to leakage and corruption are not easily available, the scheme ranked high in terms of screening efficiency (low type I errors),There was a sharp decline in the coverage of the EGS after 1988 following a virtual doubling of the wage rates in May 1988. Prior to that, EGS wage rates were less than market wages but this was reversed with the wage hike. According to some studies, this resulted in rationing of employment opportunities, leading to exclusion of eligible participants (and thus eroding the nature of the "guarantee").

In more recent years, observers have also noted a deterioration in other elements that translate the guarantee into actual delivery of the EGS benefits, including informal program guidelines, extensive monitoring, unscheduled field visits, vigilance tours by officials at various levels, and the advisory and supervisory roles of nonofficial statutory committees. Although type I targeting errors are not considered a source of major concern, the declining coverage, quality and maintenance of rural assets created and problems of governance (type II errors) are noteworthy, given the exemplary history of the EGS. For example, a recent review found wage employment generated through food for work continued to be important, contributing a significant 40-45 % of total family income of beneficiaries but the work was not organized in the lean season, and bribes had to be paid to obtain the employment, (PDI(2000)). Even more recently, a petition has been filed in the Bombay High Court accusing the state government of diverting money from the Employment Guarantee

Fund of the EGS into its general-purpose budget (Bavadam.(2003)). Ironically, the legal action has been brought by a former member of the Planning Commission who is also a member of the committee set up by the state government to review the Maharashtra Employment Guarantee Act 1977. The government believes the trend decline in coverage of the EGS reflects a declining need for the scheme in view of improvements in rural economic conditions.

TOWARDS FASTER ECONOMIC GROWTH AND POVERTY ALLEVIATION

Building local enterprise: The State is almost entirely dependent on central resources for its finance and its increasing budget expenditure is due to this rather than fast economic growth in the State. This has encouraged dependency syndrome rather than building the entrepreneurial impulse of the local population. To develop qualities of enterprise, GOI might ask the leading industrial houses, such as Reliance and Tatas, to select the most promising 100 students in the age group 15-18, train them in their companies, and employ at least 50% of them for at least 5 years. GOI should bear the entire cost, which will not come to more than 5 crores a year. This can be easily met from the existing central assistance to the state. The scheme will make Meghalaya's young generation work outside the state, will give them confidence, turn them into good company managers, and then only they would become entrepreneurs in their own home state.

Downsizing bureaucracy: Despite liberal assistance from GOI of more than 10 billion Rs, or 25,000 Rs per family per annum, state lags behind in development. The main reason for this sorry state of state finances is the huge establishment cost, which is several times state's own revenues. The number of government servants in Meghalaya increased by more than ten times since the creation of the state, although its per capita NSDP during the same period did not even double itself. The only long-term solution is to achieve adjustment through downsizing of bureaucracy. However NE states cannot do so on their own, unless GOI, Planning Commission and larger states take the initiative.

Abolish Councils: The delivery system in Meghalaya is seriously constrained because of the existence of now totally redundant Autonomous District Councils that are a burden to the tribal society. The Sixth Schedule should accordingly be amended and the state should deal directly with subjects like land and forests, and with the traditional people's institutions that are still fairly robust and have credibility.

Create land records: There are no land records in Meghalaya. The main beneficiaries of the present regulation-less situation in rural Meghalaya, especially in Khasi and Jaintia hills regions, are the upper sections, those with more economic and political power. They are able to enclose former commons and have acquired huge land holdings. It may not be possible now to go back to the community systems of managing land, as the evolution over the past few decades has been in the direction of strengthening the elements of private property, though in such a laissez faire manner that the more powerful have gained more. The only safeguard for the poor would then be to enforce land legislation. But however desirable it may be to start a land survey for the entire state, it is not going to be practically possible. A way forward is to try to institute land records and land regulations at the village level on a pilot basis first. A transparent system of land records at the village level can help secure the land rights of the poorest, to the extent that they are organised and able to articulate their interests.

Overhaul poverty alleviation schemes: Such schemes all over the country are funded from GOI funds as Centrally Sponsored Schemes, and unfortunately states have little option in changing these schemes or adapting them to suit state needs. These schemes have been in existence for more than two decades, and have run out of their potential. These suffer from inherent weaknesses and are in need of drastic overhauling.

Rather than running rigid and straight-jacketed schemes of GOI that allow for no flexibility to the states, Meghalaya government should formulate its own schemes and put pressure on the GOI to permit Rural development funds of GOI to be used for enhancing the budgetary allocation of state specific rural development schemes, or for meeting the state contribution for donor assisted programmes for poverty alleviation. Such schemes will satisfy local needs and will also lead to better people's participation.

Initiative to change the nature of these schemes cannot be taken by Meghalaya alone. It has to come from the Planning Commission and the Ministry of Rural Development. One should move away from beneficiary oriented schemes (that breed corruption) to building assets for the community. Secondly, allocation for safety net programmes that give subsidy or cash should be reduced, and assistance in terms of food should be increased.

Improve nutrition of the poor: The fund allotment for Annapurna, Old Age Pension Scheme, Mid Day Meal Scheme and ICDS should be doubled in those districts of Meghalaya where 30% or more people are below the poverty line. These measures will improve nutrition and consumption of the poor, without leakages of the magnitude seen in TPDS. The working of the Block Selection Committee in the State also needs to be improved.

Agriculture: Modest but useful unexploited potential for groundwater, river lift and small diversions offer significant immediate possibilities by actively promoting the use of affordable and cost-effective means of irrigation. A new initiative has been taken by Assam to popularise diesel pumpsets for small farmers. Meghalaya should learn from this, and also from the experience of Bangladesh and Darjeeling where manual pumpsets called treadle pumps are very popular, requiring no running cost, except family labour.

The climatic conditions of the Northeast offer good potential for horticulture and plantation like tea, rubber and spices. The productivity has to be improved in the case of pineapple etc. through tissue culture and the development of breeder seed farms. Fisheries, dairy and poultry have high growth potential in the state. But they face problems in marketing because of lack of storage facilities, non-availability of cold storage, inadequate food processing facilities and limited local market. As a result there is a lot of wastage in terms of both resource and labour. Low cost food-processing units and cold storage should be encouraged to prevent wastage and to add value.

At the moment there are too many schemes in agriculture, which dilutes the impact. The procedure for issuing financial sanctions, although simplified lately to some extent, is still dilatory. Further, the schemes assume that supply of inputs is the main constraint, whereas one should address demand related constraints. The procedure for availing various kinds of subsidy on inputs is rather complicated, as it requires several contacts of the farmer with bureaucracy.

Shifting cultivation : The whole approach should be to build upon traditional wisdom, technology and knowledge base through modern scientific inputs, based upon a value system with which the people could identify and therefore participate effectively in the developmental process.

One should consider transfer of technology from one tribe to another as one of the pathways. Strengthening the agroforestry component of the shifting agriculture system, using locally acceptable species such as the Nepalese alder, and the improvement of valley agriculture and home gardens with appropriately identified scientific inputs (linking traditional and modern technology) should also be considered. Products from jhum should be advertised and marketed as organic products.

Forestry: As already suggested, ADCs should be abolished, and people's committees and government should jointly manage non-private forests, as in other states. Rights of ownership of the individuals who now control most of the forests should be recognised, so that there is security of tenure for private forestry. Forest Department in Meghalaya has a good track record of raising plantations on private lands. The funding for this component needs to be increased. Management plans should be prepared soon, and approval of the Ministry of Environment and Forests be obtained. Forest laws and regulations should be simplified and unnecessary controls should be removed.

Suggestions for NTFP Development-The following could enhance the lucrativeness of NTFPs as a source of cash income to the rural poor:

(i) Collection, processing and marketing of NTFPs be properly organised (e.g., through self-help groups), that would leave the producers with a greater share of the value addition, and thus make investment in NTFPs more lucrative to the rural households.

(ii) The producers' access to seeds and seedlings of good genetic quality should be improved.

(iii) The producers need assistance in establishing market connections with a minimum of middlemen.

(iv) Medicinal and aromatic plants for export could be grown under contracts with domestic firms to eliminate too many middlemen.

(v) A simple market price information system easily accessible by the rural communities should be established.

(vi) Microcredit facilities should be accessible to the rural households/ communities for investing into simple processing equipment and facilities.

Delivery-Although problems faced by the northeast are far more serious than existing in other poor states of India, institutional capacity to formulate schemes and implement them effectively is still to be developed. Effective implementation of suggestions described in this chapter will require bureaucracy that is committed to tribal and people's welfare and has the necessary expertise to translate pro-poor policies into action. Therefore, in addition to implementing the policy reforms detailed here, several measures will need to be taken in order to enhance the efficiency and professionalism of public service and institutions of governance.

CONCEPTUAL ISSUES

The central focus of this chapter is to study how factors external to agriculture impinge upon its performance and how such factors have affected rural poverty. We emphasize one mechanism that having potentially a large policy connotation has, in our view, been insufficiently emphasized and analysed by the literature: The way in which the public sector spends tax revenues in the rural sector.

We do not intend to evaluate whether or not governments spend enough in agriculture. This is an old debate. We instead are interested on the structure or composition of government expenditures in rural areas (including of course in agriculture). But we are not interested either on analysing the impact of particular government expenditure items.

The literature is full of studies that do this. Measuring the impact of road expenditures, R&D investments, credit subsidies, farm extension, rural education, expenditures in rural health, are some of the favourites in the literature. This chapter makes a contribution by showing that certain specific items may have high pay-offs while others do not. These findings may have relevance by emphasizing the need to improve the cost-benefit evaluation of particular public expenditures and investments. But it is difficult to obtain more general policy implications from them.

We instead use a broader taxonomy of public expenditures that is conceptually and policy grounded. The classification of public expenditures into public (or semi public) goods and private goods arises from the conceptual view that governments' prime responsibility is to use taxpayers money in supplying public goods, including goods and services that are under supplied as a consequence of missing markets or of market imperfections. To be sure, there are many grey areas. Sometimes expenditures on private goods are difficult to distinguish from expenditures in public goods. But by and large these are the exceptions; in practice more often than not the separation is quite sharp.

Political motivations and socioeconomic consequences of public expenditures. The source and political motivation of the government's decision to invest in public or private goods is fundamentally different and public/private good choices have important socioeconomic consequences. By definition the benefits of public or semi-public goods tend to disperse much more among the population than those of private goods. Thus, political lobby groups agitate for more private goods that wholly benefit their particular constituency group than they do for public goods of which the lobby group shares only a fraction of the benefits. The struggle by lobby groups to elicit subsidies from the government is often their prime motivation. It is well known that most successful lobbies are typically small groups with a clear common interest and objective (Olson, 1965). In addition, and most importantly, successful lobby groups almost always have enough financial means to bribe and to pay political campaign contributions to politicians, or have enough social status to be able to affect public opinion through a variety of means.

The above requirements for constituting a successful lobby group are often satisfied by economic elites and rarely by the poor. The implication is by now a stylized fact: government expenditures in subsidies and other private goods go almost always to the wealthy, only very rarely to the poor. This is in sharp contrast with expenditures in public goods, which generally produce benefits that tend to disperse across all groups in society, poor and non-poor. The social equity effects of the composition of public expenditures are obvious. The greater the share of government expenditures in private goods the worse, ceteris paribus, is income distribution and poverty. Because the structure of public expenditures tends to be more biased towards private goods when the economic power of the elites is in the first place already greater, such biased composition of public expenditures is both cause and effect of poverty and social inequity.

The structure of public expenditures and economic efficiency. The public/private good dichot-omy is grounded on another factor which has to do more directly with economic efficiency than with equity. Investment in public goods provides factors of production that are almost by definition rarely supplied by the private sector. For this reason one can reasonably expect that such investment will fill a vacuum that is not fulfilled by the private sector. As a consequence, a complementary relationship between private investments and public goods is likely to arise. Government expenditures in private goods (subsidies), by contrast, often finance assets that can equally be financed directly by private entrepreneurs. Even more importantly, as argued above these government subsidies go to the wealthy, those that face generally no constraints to invest at optimal levels (that are likely to face no liquidity constraints or other types of market imperfections). As shown below, the effect of these subsidies is almost always detrimental for economic efficiency and, far from contributing to promote private investment, they tend to crowd it out and to increase the consumption of the wealthy instead. Thus we have another important contrast that justifies the public/private goods dichotomy. While increasing public goods is likely to promote economic growth directly (as factors of production) and indirectly (through its positive effect on private investment), increasing subsidies is more often than not deleterious for growth and private investment. Crowding-out and agricultural performance. Government provision of private goods or subsidies causes three forms of crowding-out with negative efficiency and equity effects:

(i) Subsidies crowd out the supply of public goods through the government budget, institutional and human constraints. This first crowding-out is obvious; the government's efforts to provide subsidies compete with the provision of public goods, not only for government's financial resources but also for other public resources that also tend to be scarce. The administration of subsidy programs often absorbs a large share of the limited supply of the government's human capital. Scarce institutional capital is also preferentially allocated to such programs. This means that subsidies directly crowd-out financial and non-financial government efforts to provide public goods. The crowding out of public good expenditures within the government's resource budget can have large negative effects on social welfare. Governments forego investments in public goods that according to many studies have extremely large rates of return. The survey by Alston et.al.(2000), for example, reviews hundreds of studies around the world estimating rates of return to agriculture R&D revealing extremely high and non-declining rates. Psacharopolous's (1994) survey does the same for studies measuring returns to education over many countries, also showing very high rates. Similarly several studies at IFPRI by Fan and his associates (2000) have found unusually high returns to a variety of public infrastructure in agriculture in India and China. The World Bank (2000) has gathered evidence pointing to also very high rates of return to investments in certain investment in protecting the environment. Thus, on the basis of this rather massive evidence one may presume that governments under invest in public goods (the persistence of such unusually large rates of return can only be explained by insufficient investment) and that such under investment is likely to have large negative effects on welfare. This conclusion is reinforced by further empirical evidence showing that subsidies have low or even negative economic rates of return and that they crowd out private investment. We turn to these issues next.

(ii) Subsidies may directly crowd out private investments. Consider what is often regarded as a "desirable" subsidy; the government offers to pay a portion of the costs of a particular investment. Assume further the best possible circumstance in terms of the allocation of the subsidy. The subsidy is of course rationed as the funds are obviously less than the demand, but their allocation among producers is transparent, not subject to corruption. Consider an investor that is able to extract a profitable return out of an investment (even in the absence of the subsidy) that potentially may qualify for the subsidy. Suppose that in that year there was a large demand for the subsidy and that the investor was not lucky enough to get the subsidy. The producer may go ahead with the investment (and never get the subsidy) anyway since it is a profitable one. Alternatively, she/he may opt to postpone the investment and try again next year in the hope of then getting the subsidy. If the expected value of the subsidy is sufficiently large to compensate the foregone profits in one year the producer may decide to delay the investment. Thus, investments that are privately (and socially) profitable may be postponed as a consequence of the existence of the subsidy. Among the investors that actually get the subsidy there are two types: Those that would have implemented the investment anyways and those that would have not (because they would not be able to get high enough returns) but they in fact invest because they have obtained the subsidy. For the former the subsidy was ineffective, the subsidy is likely to promote more consumption by them rather than more investment as intended. For the latter the subsidy was effective in causing them to invest but, in the absence of positive externalities associated with the investment, at a low social return. Thus, the subsidy scheme does two things. First, it increases consumption of producers that would have invested anyway but that are able to obtain the subsidy. Second, a reallocation of investment from producers that potentially can get high rates of return but do not obtain the subsidy (and decide to postpone the investment expecting to get the subsidy in the future) to producers that obtain a low social return but invest only because they are able to access the subsidy. The net effect on total investment is ambiguous, but the efficiency impact is negative as a consequence of the second effect. Moreover, subsidies induce agricultural growth to be more based on extending. The above example is not just a curiosity. It illustrates a phenomenon that has received important empirical support in recent years. That subsidies, at least in the form in which they are usually allocated, do not generally promote investment or more R&D has been shown by several studies in various countries. Empirical studies using detailed firm level data by Bregman et.al. (1999) for Israel, Fakin (1995) for Poland, Lee (1996) for Korea, Bergstrom (1998) for Sweden, Estache and Gasper (1995) for Brazil, Harris (1991) for Ireland and several others have shown that subsidies and corporate tax concessions are at best ineffective to promote investment and technological adoption and, in some instances, even counterproductive. Crowding out of private investment as a consequence of the subsidies occurs. The land area than on intensifying production

(iii) Subsidies may indirectly crowd out private investment in the intermediate run. The third type of crowding-out of private investment due to public expenditure biases toward subsidies is long-run in nature. The low stocks of public goods caused by continuous under investment in public goods, contributes to low productivity and diminished margins of return for private investments over time. This, in turn, is translated into lower future levels of private investments, lower productivity, and lower growth over the long run.

The structure of Public expenditure and rural poverty. Public goods affect rural poverty through a direct effect and through an indirect effect:

(i) Direct effects. Important components of rural public goods are human capital and investment in the protection of natural resources and the environment. Human capital is the main productive asset for most poor. However, as a consequence of credit and other market failures they are not in general able to fully finance investments in human capital, regardless of how high the rate of return to these investments might be (World Bank, 2000). They are, therefore, largely dependent on the public sector as a source of financing for these investments. On the other hand the rural poor are also highly dependent on natural resources as a source of subsistence (Barbier, 2004). Resource degradation is often disproportionly paid by the rural poor. Increasing investment to protect natural resources and in mitigating environmental externalities is, therefore, vital to reduce rural poverty.

(ii) Indirect effects. As discussed earlier an adequate stock of rural public goods contributes to sustain growth for agriculture and related rural industries. In general much of the rural industry linked to agriculture is highly intensive in unskilled labour (López and Anriquez, 2003). Moreover, it appears that the rural sector in many developing countries is important enough within the unskilled labour market to significantly affect the real wages for unskilled workers even at a national level. Thus a dynamic rural productive sector is a vital source of employment for unskilled workers as may play a role in supporting higher real wages. To the extent that most poor are unskilled workers, these effects may be important in reducing poverty.

CONCLUSIONS

This chapter has provided evidence suggesting that the structure of public expenditures is an important factor of economic development. The quantitative importance of this factor appears to be greater than the traditional factors on which the development literature focuses. In particular, expanding total public expenditures in rural areas while maintaining the existing public expenditure composition prevailing in the countries does little to promote agricultural income and reduce rural poverty. The key issue is not so much how much money is spent in the sector but rather how are public monies being spent. Spending a significant share of government resources in subsidise to the wealthy causes not only less agriculture income but it also induces an excessive reliance of agricultural growth on land expansion, thus exacerbating the negative effects of agriculture on the remaining forests.

The econometric results provide estimates that suggest that governments over spend in mostly politically-motivated private goods and subsidies and under spend in public goods and anti-poverty programs. Though the very large estimated effects of expenditure composition may be surprising to some, one should note that these estimates are highly consistent with two previous pieces of evidence available in the literature.

First, the literature reports extremely large rates of return to many forms of public goods (often 25% to 60% per annum and above) including R&D, education, health care, some infrastructure, certain environmental protection investments, and others. Also, those studies that have performed analyses over time generally find that the rates of return show no tendency to decline over time.

Recent surveys of this literature, cited earlier in the text, find an amazing degree of consensus on the estimated rates of return obtained in many studies that evaluate rates of return to public goods around the world. The fact that the studies reviewed have used different approaches, diverse data bases and have been applied to a large number of countries around the world, means that this consensus should be taken seriously.

These extremely high and non-declining rates of return suggest that there are important investment opportunities in public goods that are not being fully exploited. That is, it suggests under spending on public goods.

Second, many firm level studies around the world (some cited earlier in the text) have found that subsidies aimed at promoting investment, employment, and growth do not achieve these objectives. In some cases they are counterproductive to these goals. Thus, while the rates of return to public goods are very high, the rates of return to private goods or subsidies are very low or even negative. Yet, governments tend to spend a large share of their resources in the goods that have low returns while at the same time forego investments with very large rates of return. Switch spending priorities and you can obtain massive dividends. This is what the econometric results are saying.

As discussed earlier in the paper, governments make such an aberrant choice mostly because they respond to a variety of politico-economic forces. These forces at play are not only associated with openly corrupt practices, but they may also reflect a highly unbalanced configuration of political forces. The elites, because of their wealth and "contacts", are able to influence even well-meaning governments to their own advantage even if the rest of society suffer loses.

Apart from political campaign contributions, the elites are able to shape public opinion, mainly through the presentation of intellectual arguments and adroit use of the media, to favour or to accept such flagrant misallocation of public resources. International institutions can, if they want it, be important players in creating domestic conditions favouring correcting such misallocation. They can also use their influence on the countries to reduce rather than deepen the massive imbalances in lobbying abilities between the elites and the rest of the civil society.

8

Agro-processing Development and Supply Chain Management

AGRO-PROCESSING EQUIPMENT SUPPLY CHAIN

Historically, formal manufacturing, distribution and sales of agro-processing equipment have been a preserve of conglomerate companies such as Precision Grinders, G. North and Son (Pvt) Ltd and Zimplow Limited. Processing systems were split into manual tools and machines for small-scale users (mainly for communal farmers) and motorised equipment for large-scale processors.

Since the mid 1990s, many new small-to medium-scale formal and informal manufacturers entered the market resulting in a dramatic increase in medium-scale motorised systems for processing food crops. It is now possible at medium-scale level to process cereals into various flours and stock feeds, groundnuts into peanut butter, sunflower, cotton and soybeans into edible oil and fruits and vegetables into jams, pulps, juices, pastes, sauces, pickles, and confectionery products.

The range of processing equipment is limited to manual and motorised shellers/threshers for cereals and pulses. Most of the manufacturing activities are located in Harare, Norton and Bulawayo with distribution networks in the major cities, towns and Growth Points or Rural Service Centres (RSCs). The local industries have the capacity to manufacture complete shellers/threshers. However, for motorised equipment, the sources of power (electric motors and engines) have to be imported from South Africa and/or Asia. This obviously has a negative bearing on the availability and final cost of motorised equipment as foreign currency supplies in the country are limited. The major consumers of primary processing equipment are individual farmers or farmer groups and private contractors.

The growth of secondary agro-processing in Zimbabwe is quite evident in the rapid rise in numbers of equipment manufacturing companies, dramatic increase in processing enterprises and the widespread availability of various processed foods on both the formal and informal market. Growth has been greater in the small-and medium-scale enterprises than the large-scale sector; for example there has been a huge increase in millers of maize and other cereals, processors of peanut butter, manufacturers of livestock feeds and soybean products.

There is clear evidence of widespread activities in manufacturing of cereal flour, peanut butter and animal feeds as indicated by the increased number of manufacturers of related equipment. This development should in theory lead to a wider choice of equipment and a level of competition that would benefit the consumers (Nazare, 2005). It is however, important to point out that while there are many players in the manufacturing sector; variety in equipment design is very limited. Most of the manufacturers rely heavily on procuring equipment and/or components of equipment from Asia as well as the local formal and informal outlets. This approach may result in poor equipment quality control and certainly results in limited consumer choice. Ultimately, there are also costs associated with fixing poor quality machines when they break down. Nonetheless, the positive side of it is that it reduces the cost of tooling for individual manufacturers, may guarantee interchangeability of spares among equipment from different manufacturers and creates employment for the locals.

Research and Development

Development of agro-processing equipment has not received the level of research support it deserves. There are limited and isolated research efforts in the private sector. In the public sector, budgets for Research and Development have been cut down drastically and furthermore there is a critical shortage of qualified research staff to carry out the work. However, considerable research in the development of crop shellers, threshers, vegetable oil mills, peanut butter mills and crop dryers has been conducted by the Ministry of Agriculture (MoA) through the Institute of Agricultural Engineering, the University of Zimbabwe (UZ) and ENDA Zimbabwe since 1990. Useful research results have been generated but extension of the results to the private sector (manufacturers) and farmers has generally been poor.

UZ usually works in partnership with the private sector in equipment manufacturing. Transfer of technologies to end-users is facilitated by the linkages that exist between the university and the public sector and NGOs.

Involvement of students in various research activities also helps in the dissemination of research results.

The development and extension of the peanut butter technology has been a major success story for the Development Technology Centre (DTC) at the UZ. Technology manufacturing and marketing is now fully market-driven. However, there are issues surrounding intellectual property rights and patenting which are not in place. Many a time, technology generated by academic or public sector research organisations ends up being used for commercial purposes without due recognition of the people involved in knowledge development. There is no evidence of real budgets for research and development in the private sector save for Precision Grinders, Hastt Zimbabwe and Practical Action that have established research and development divisions. Others depend on technology importation (plus adaptations) and research results from local work.

Extension and Training

Training in agro-processing is at three levels; manufacturers, distributors and end-users who are either farmers or entrepreneurs. Training in manufacturing techniques is not a limitation in Zimbabwe. Training of artisans (tool makers, welders, fitter and turners, sheet metal workers etc) is well established and of high standard. Training Institutions such as Silveira House, Glen Forest, Hlekweni Friends, the Institute of Agricultural Engineering of MoA offer training in blacksmithing targeted at rural artisans. Rural blacksmiths carry out minor repairs on equipment at local level. However, most of these institutions are operating below normal capacity due to financial constraints or high staff turn-over.

There is very limited training offered to distributors as basic knowledge about the operations of equipment is frequently missing at points of sale. This is often exacerbated by lack of operational manuals on some products. Where attempts are made to provide manuals, they sometimes lack necessary assembling and trouble-shooting details.

Very few small-and medium-scale agro-processors have received formal training in food processing techniques. Most of the formal training support for small-scale food processors has been provided by Ranche House College with funds provided under a Dutch-funded programme called "Food Processing as a Small Business". The training included solar drying of fruits and vegetables, technical and business skills, and entrepreneurship and financial access and management.

AGRICULTURAL MARKETING AND AGRI-BUSINESS DEPARTMENT

State Scheme

Own building for Agmark Laboratories, Strengthening of Agmark Laboratories and Provision of Computers (₹ 3.07 crores): There are 30 State Agmark Grading Laboratories, one Principal Laboratory and 15 Agricultural marketing centres functioning in the State.

It is proposed to construct building for Agmark Labs as well as modernize the equipments during the Tenth Five Year Plan for which a sum of ₹3.07 crores is provided.

Schemes Proposed to be Implemented Availing Financial Assistance from the other Financial Institutions

Provision of Infrastructure Facilities (₹ 100 crores)

i. *Creation of infrastructure facilities:* There is a need to provide infrastructure facilities like transaction shed, drying yard, farmers rest sheds, sanitation facilities, drinking water supply electronic weighing scales, cleaning and washing facility, moisture meters, scientific instruments for grading, rural godowns etc., for which an amount of ₹ 100 crores is provided.

ii. *Revamping Tamil Nadu Agricultural Marketing Board:* There is an imperative need to revitalize the Board activities by making this body more responsive and independent decision- making body. It should be given administrative and financial powers.

iii. *Revitalising the market committees:* At present the Act provides for nomination of members to the Market Committees. To make it more accountable and democratic, it is necessary that members be elected so that there is a sense of belonging. Secondly, there is lack of professionalism in the staff of the Market Committees. Their mind set is more of regulation than market oriented approach. Training to the existing staff and inclusion of professionals is a must.

iv. *Provision of infrastructure facilities in Post Harvest Centres:* At present the post harvest centres in the Regulated markets conduct training programmes to the farmers and exhibitions on various post harvest practices. The regulated markets should expand their activities through provision of whole storage CA, MA, Retardation and ripening facilities for the agricultural produce.

v. *Agricultural Marketing Extension:* Strong network of marketing extension is very much necessary at block level to effectively advise farmers on various aspects of marketing, advise on product planning, marketing information, and securing market for farmers, and advise on improved market practices and advise on post harvest management practices.

To strengthen the regulated market it is proposed to provide infrastructure facilities like transaction shed, drying yard, farmers resting shed, sanitation facilities, drinking water supply, electronic weighing scales, cleaning and washing facility, moisture meters, scientific instruments for grading, rural godowns etc. at a cost of ₹100 crores with financial assistance from Marketing Committee, NABARD, GOI.

Provision of infrastructure facilities in Post Harvest Centres-Cold Storage (₹ 200 crores- National Horticultural Board/Private Sector)

It is estimated that around 30% of the horticulture produce is wasted due to inadequate cold chain facility and appropriate technology for the preservation of horticultural produce.

At present the combined cold storage capacity of 133 units in Tamil Nadu is around 1 LMT. The existing capacity is not sufficient to store horticultural, dairy, and marine products. In order to meet the demand it is proposed to establish an additional 1 LMT capacity of cold storage facility, at an estimated cost of ₹ 200 crore. Taking into consideration the initial capital investment, high recurring cost and low capacity utilisation and project failure, it has been proposed to attract investment by private sector through measures like financial incentives in addition to NHB's subsidy and power tariff concession.

Agricultural Marketing Extension/ Packaging Training (₹ 42.95 crores)- Tamil Nadu State Agricultural Marketing Board (TNSAMB)

Agricultural farmers require advice on various aspects of marketing like selection of the crops to be grown with marketability in mind, current price, market arrival and forecasting of market trends, and on post harvest management practices.

Agro-processing Food Park (₹ 10 crores)- Private Sector

It is estimated that about 10-20% of the food grain production and 30-35% of fruits and vegetables are wasted at various stages from picking to consumption. This indicates the need to establish food-processing industries to preserve and minimize the wastage. To encourage private sector to establish agro-processing industries, the Government of India, Ministry of Food Processing Industries has formulated two models of Food Park for Food Processing Industries apart from other incentives. It is proposed to set up a

Food Park at a cost of ₹ 10 crores either in private sector or by Market Committees.

Setting up of Terminal Markets and Collection Centres (₹ 160 crores)

Private sector or growers association in partnership with private sector can organise Terminal Markets for specified products with backward integration with collection centres (value addition centres). Fifteen to twenty collection centres established nearer to the production area, can feed one Terminal Market. Two such a Terminal Markets will be established one each at Chennai and at Coimbatore with NDDB assistance.

Mega Markets (₹ 10 crores)

A mega market centre can cater to the needs of wholesale dealers, exporters and food processing industries. The availability of horticultural products in large scale in one place would promote exports and provide single sourcing to agri processing industries. It is proposed to establish a mega market for vegetables at Oddanchatram in Dindigul District at a cost of ₹ 2.59 crore. Depending on the success of this project, mega markets could be developed at Mettupalayam in Coimbatore District and Athur in Salam District either through Market Committees or private sector.

Agro Processing Industries (₹ 10 crores)

Food Processing Industries provide the crucial farm - industry linkages, which help to add value to the produce, generate employment opportunities and increase the net income to the farmers. The Market Committees will establish these through private sector in 10 places with an incentive of 20% of the capital cost. The project cost is estimated to be ₹ 10 crores. Packaging fresh fruits, vegetables and other farm produce is an important process as it reduces post harvest losses, increases the income of the farmers and ensures clean and hygienic farm produce to the consumers. It is proposed to create awareness and impart training on farm produce packaging to the technical officers of Agriculture, Horticulture, Agricultural Marketing and Agri Business departments during the Tenth Plan period. It is proposed to train 600 officials and 400 farmers at a total estimated cost of ₹ 25 lakhs. It is also proposed to conduct studies for developing low cost packaging techniques and material for some specific commodities at a cost of ₹ 0.35 crore.

Agri Export Zone (₹ 150 crores) –GOI

The most critical factor to meet the challenges of export will be enhancing exporting capability of our State in a highly competitive environment. To promote export of agri products, Government of India, Ministry of Commerce has announced in 2001 - Exim Policy, a scheme of establishing Agri Export Zone (AEZ). In AEZ, institutional and physical infrastructure would be created as per the needs of the specific commodity. Steps have been taken to establish AEZ for cut flowers, mangoes, bananas, medicinal plants and vegetables. To promote agricultural exports, it is proposed to educate and train the growers of identified crops in producing, grading and packing for international market, and in establishment of analytical laboratories, setting up of Export promotion cell in Agri Business department to disseminate information and harmonization of standards of Indian products with international standards (Codex).

Marketing Extension

Network is proposed to be formed integrating with the extension network already available with Agriculture Department. Officers of Agriculture, Horticulture and Agricultural Marketing departments will be given training on various aspects of Agricultural Production and Marketing for the purpose of carrying out extension works effectively and efficiently.

A sum of ₹65 lakhs is proposed during the Tenth Five Year Plan. Further, the Government will also facilitate private sector to carryout extension work for the quick reach of farm information to the farmers.

To improve market practices and also sale of specific agricultural products, two products' specific market Complexes *viz*- Turmeric Market Complex at Erode at a total cost of ₹32.30 crores, and Jaggery Market Complex at Trichy at a total cost of ₹10 crores will be established. These market complexes will be established incurring expenditure initially from Market Committees and financial institutions like NABARD and then recovered from the traders in instalments.

Alternative Marketing Forms

Role of Government in managing markets is on decline worldwide. It is not easy to bring major changes in the traditional marketing system. The only way to modernize marketing is to promote alternative marketing system and that may operate parallel to and in addition to present marketing system. The purpose of the proposed alternative marketing is to promote modern trade practices, which in turn will pave way for transperancy and effiecncy in market.

Even though, the various forms of alternate marketing like:

- Direct marketing,

- Marketing through farmers interest group,
- Setting up of terminal markets,
- Forward and future market,
- e-commerce,
- Setting up of mega markets,
- Negotiable warehouse receipt system etc. have been suggested by Expert Committee on Agricultural Marketing headed by Shankarlal Guru, three important marketing methods could be considered in the State *viz.*, Terminal Market, Mega Market and Direct Market.

Agro Processing Industries

Food Processing Industries provide the crucial farm - industry linkages which helps to accelerate overall agricultural development, adding value to the produce, generating employment opportunities and increasing the net income to the farmers. In India, only 2% of the total horticultural produce is processed. In countries like Brazil, it is in the range of 70%. Considering the rising demand for good quality products, there is an urgent need to enhance capacities for value added and processed products. At present, value addition is estimated at 7% of the total production within next 5 years.

There is a need to increase value addition to 20% and processing at 7%. To increase our country's share in the world trade of agri products which stands at less than 1% at present, the most critical factor in the highly competitive market environment is quality processed products. In Tamil Nadu, food processing in the form of drying, vegetable oil, grain processing, sugar breweries are in existence for a quite long period. Lack of adequate infrastructure facilities like storage, processing, marketing besides technical know-how have been the major constraints affecting the growth of the industry. Tamil Nadu with a coastline of 922 km and surface boundary of 1200 km with tropical and sub-tropical climate is ideally suited for the production of a host of agricultural, horticultural, acquacultural and animal husbandry produces. Thus there is vast scope for setting up of Food Processing Industries in Tamil Nadu.

The growth of food processing industry, which is included in the priority lending sector, will bring immense benefit to the economy. Economic liberalization and raising consumer prosperity is opening up new vistas in food processing sector. Tamil Nadu produces 103 lakh metric tonnes of food grains, 22 lakh metric tonnes of oil seeds, 37 lakh metric tonnes of sugarcane, 53.89 lakh metric tonnes of fruits, 54.65 lakh metric tonnes of vegetables, 2.95 lakh metric tonnes of spices and 6.94 lakh metric tonnes of plantation crops. 10 to 20% of food grain production is lost every year both at pre and post harvest stages. Similarly 30 to 35% of fruits and vegetables produced are also wasted at various stages from picking to final consumption. This only indicates the immense potential and scope that exists for setting up of food processing industries so that the available raw materials may be put into the most judicious use. This also indicates the scope that exists for adoption of post harvest modern techniques ensuring preservation and minimizing wastages of agricultural produce.

OPPORTUNITIES IN THE AGRO-PROCESSING INDUSTRY DEVELOPMENT

Agro-processing opportunities in Zimbabwe currently tend to favour growth and development of medium-scale processing industries that match the current production levels and the distortions in marketing of produce. Market forces and the prevailing economic environment favour more down-sizing of large-scale processing systems and upgrading small-scale processing industries. This is mainly due to the fact that the demand for raw materials by large-scale manufacturers is currently not being met due to low national production levels hence the enterprises are operating below capacity. This has resulted in scaling down of business, massive staff retrenchments and/or closure of factories. For example Aroma Bakeries in Harare sold their peanut butter business (equipment and franchise) to a medium-scale processor. Some large-scale operators contemplate sub-contracting small and medium processors who can meet their set standards. The contractor would provide packaging materials and take responsibilities for transporting and marketing the product. This arrangement shifts raw material sourcing and processing risks away from the large-scale processor.

Small-scale enterprises which are characterised by use of manual equipment are losing business to the medium-scale as the latter use motorised equipment which have advantages of both higher capacity and efficiency. Manual equipment is either abandoned or upgraded to match the demand. Examples are: (i) the rapid obsolescence of manual sunflower oil presses in the 1990s following the importation of motorised multi-crop oil expellers from Asia and (ii) the current requests by peanut butter processors to upgrade manual mills by installing motors. Small-scale processors who cannot afford to improve their processing systems are likely to get out of business.

There is great potential in the development of medium-scale grain milling, bread-making enterprises, livestock feed manufacturing, peanut butter and vegetable oil processing as opportunities of market entry by medium-scale entrepreneurs increase. Fruit and vegetable processing and preservation offer a new viable opportunity though more effort needs to be put into product promotion and marketing.

For example Lamin and Son in Nyanga started processing a wide range of fruits as a pastime but later expanded into a medium-scale factory capable of producing half a tonne of dried products per day for both domestic and export markets.

There are widespread shortages of fruits and vegetables in the drier parts of the country yet during peak production periods there is a glut and large amounts of fruits and vegetables are put to waste. Fruits can be processed into pulps/juices, wine, jams, jelly, marmalade, pickles, dried products and confectionery items.

Small and medium-scale agro-processing industries at RSCs are strategically located to take advantage of both raw material and local markets. The dominant activities in RSCs are cereal milling, peanut butter milling, and vegetable oil processing. There are also opportunities for fruit and vegetable drying, canning, bottling, juicing and freezing in the RSCs (Price Water House, 1994). Equipment required for small-scale fruit processing is based on the usual home kitchen gadgets including stoves, pots, knives and spoons which are widely available in hardware shops around the country.

For medium-scale enterprises, more specialised equipment such as stainless industrial boilers, pulpers, pasteurizers and canning machines become necessary. These are again available from Stainless Steel Industries and Ag-Venture respectively. The Development Technology Centre of the University of Zimbabwe reported limited awareness of the Zimbabwean market to dried fruit products as a constraint to the uptake of solar dryers.

Bibliography

Aplet, Gregory H.; 1993. *Defining Sustainable Forestry.* Washington, D.C.: Island Press.

Bains, William: *Biotechnology: From A to Z,* New York, Oxford University Press, 1998.

Barnum, Susan R.: *Biotechnology: An Introduction,* Belmont, Thomson/Brooks/Cole, 2005.

Bernauer, Thomas: *Genes, Trade, and Regulation: The Seeds of Conflict in Food Biotechnology,* Princeton, Princeton University. 2003.

Brandwein, P.F.: *Sourcebook for the Biological Sciences,* San Diego, Harcourt Brace JOvanovich, 1986.

Broach, J.R.: *The Molecular Biology of the Yeast,* Cold Spring Harbor, Cold Spring Harbor Laboratory, 1981.

Chrispeels, Maarten : *Plants, Genes and Crop Biotechnology,* Sudbury MA, Jones and Barlett Publishers, 2003.

Clark, J.M.: *Experimental Biochemistry,* New York, W.H. Freeman and Company, 1977.

Conko, Gregory: *The Frankenfood Myth: How Protest and Politics Threaten the Biotech Revolution,* New York, Praeger, 2004.

David Sadava: *Plants, Genes and Crop Biotechnology,* Sudbury MA, Jones and Barlett Publishers, 2003.

Davis, Thomas. 2000. *Sustaining the Forest, The People, and the Spirit.* Albany, NY: State University of New York Press.

Dixon, Dougal: *After Man-A Zoology of the Future,* New York, St. Martin's Press, 1981.

Edwards, C.A. *Sustainable Agricultural Systems.* 1990. Ankeny, Iowa: Soil and Water Conservation Society.

Franklin, Steven E. 2001. *Remote Sensing for Sustainable Forest Management.* Boca Raton, Fla.: Lewis.

Fransman M, Junne G, Roobeek A: *The Biotechnology Revolution?,* Oxford, Blackwell, 1995.

Fukuyama F.: *Our Posthuman Future: Consequences of the Biotechnology Revolution,* NewYork, Farrar, Straus and Giroux, 2002.

Fumento, Michael: *Bioevolution: How Biotechnology is Changing Our World,* San Francisco, Encounter Books, 2003.

Gadow, Klaus von and Pukkala, Timo. 2000. *Sustainable Forest Management.* Dordrecht; Boston: Kluwer Academic Publishers.

Gaskell G: *Biotechnology-the Making of a Global Controversy,* Cambridge, Cambridge University Press, 2002.

Goodsell, David S.: *Bionanotechnology: Lessons From Nature,* Hoboken, Wiley-Liss, 2004.

Hatfield, Jerry L. and Karlen, Douglas L. 1994. *Sustainable Agriculture Systems.* Boca Raton: Lewis Publishers.

Horn, Toby M.: *Working with DNA and Bacteria in Precollege Science Classrooms,* Reston, National Association of Biology Teachers, 1993.

Jordan, Carl F. 1998. *Working with Nature: Resource Management for Sustainability.* Australia: Harwood Academic Publishers.

Lal, R., 1998. *Soil Quality and Agricultural Sustainability.* Chelsea, Mich.: Ann Arbor Press.

Mater, Jean. 1997. *Reinventing the Forest Industry.* Wilsonville, Or.: GreenTree Press.

Metting, Jr., F.B.: *Soil Microbial Ecology: Applications in Agricultural and Environmental Management,* Marcel Dekker, New York, 1992.

Miller, Henry I. : *The Frankenfood Myth: How Protest and Politics Threaten the Biotech Revolution,* New York, Praeger, 2004.

Morholt, E. : *Sourcebook for the Biological Sciences,* San Diego, Harcourt Brace JOvanovich, 1986.

Murray, David: *Seeds of Concern: The Genetic Manipulation of Plants,* Sydney, University of New South Wales, 2003.

Murray, David: *Seeds of Concern: The Genetic Manipulation of Plants,* Sydney, University of New South Wales, 2003.

Pilarski, Michael, 1994. *Restoration Forestry: an International Guide to Sustainable Forestry Practices.* Durango, Colo.: Kivaki Press.

Poindexter, J.S.: *Strategies and Mechanisms for Field Research in Environmental Bioremediation,* American Academy of Microbiology, Washington, D.C., 1994.

Prasad, Rajendra and Power, James F. 1997. *Soil Fertility Management for Sustainable Agriculture.* Boca Raton: CRC/Lewis Publishers.

Primrose, S.B.: *Principles of Gene Manipulation,* London, Blackwell Scientific Publications, 1989.

Rees, R. M., ed. 2001. *Sustainable Management of Soil Organic Matter.* Oxon, UK; New York: CABI Pub.

Retzer, W.J.: *Biotechnology Workbook,* Englewood Cliffs, Prentice Hall, 1991.

Sayler, G.S., Fox, R., and Blackburn, J.W.: *Environmental Biotechnology for Waste Treatment,* Plenum Press, New York, 1991.

Shetty, Kalidas: *Food Biotechnology,* New York, Dekker/CRC Press, 2005.

Slater, R.J.: *Experiments in Molecular Biology,* Clifton, Humana Press, 1986.

Smith, John E.: *Biotechnology,* Cambridge, Cambridge University Press, 2004.

Switzer, R.L.: *Experimental Biochemistry,* New York, W.H. Freeman and Company, 1977.

Tollefson, Chris, ed. 1998. *The Wealth of Forests: Markets, Regulation, and Sustainable Forestry.* Vancouver: UBC Press.

Towle, Albert: *Modern Biology,* Austin, Holt, Rinehart and Winston, 1988.

Walden, Richard: *Genetic Transformation in Plants,* England, Open University Press, 1988.

Index

❑❑❑